PRIVACY AND IDENTITY PROTECTION

INFORMATION SECURITY

CYBERATTACKS, DATA BREACHES AND SECURITY CONTROLS

PRIVACY AND IDENTITY PROTECTION

Additional books and e-books in this series can be found on Nova's website under the Series tab.

PRIVACY AND IDENTITY PROTECTION

INFORMATION SECURITY

CYBERATTACKS, DATA BREACHES AND SECURITY CONTROLS

PAUL SCHOLZ
EDITOR

snova
New York

Library of Congress Cataloging-in-Publication Data

ISBN: 978-1-53615-880-9

Published by Nova Science Publishers, Inc. † New York

CONTENTS

PREFACE

Federal agencies and our nation's critical infrastructures, such as communications and financial services, are dependent on information technology systems and electronic data to carry out operations and to process, maintain, and report essential information. Yet, cyber-based intrusions and attacks on federal and nonfederal systems have become not only more numerous and diverse, but also more damaging and disruptive as discussed in chapter 1.

The IRS has a demanding responsibility to collect taxes, process tax returns, and enforce the nation's tax laws. It relies extensively on computerized systems to support its financial and mission-related operations and on information security controls to protect the sensitive financial and taxpayer information that reside on those systems. As part of its audit of IRS's fiscal year 2017 and 2016 financial statements, GAO assessed whether controls over financial and tax processing systems were effective in ensuring the confidentiality, integrity, and availability of financial and sensitive taxpayer information as reported in chapter 2.

Reliance on a global supply chain introduces multiple risks to federal information systems. Chapter 3 highlights information security risks associated with the supply chains used by federal agencies to procure IT systems.

The Office of Personnel Management (OPM) collects and maintains personal data on millions of individuals, including data related to security clearance investigations. In June 2015, OPM reported that an intrusion into its systems had affected the personnel records of about 4.2 million current and former federal employees. Then, in July 2015, the agency reported that a separate but related incident had compromised its systems and the files related to background investigations for 21.5 million individuals. From February 2015 through August 2017, multiple reviews of OPM's information security were conducted. Four reports based on these reviews were issued. The reports contained 80 recommendations for improving the agency's security posture. Chapter 4 reviews relevant documents and artifacts reflecting OPM's actions and progress toward implementing the 80 recommendations contained in the four reports, and assessed the actions against the intent of the recommendations.

CDC is responsible for detecting and responding to emerging health threats and controlling dangerous substances. In carrying out its mission, CDC relies on information technology systems to receive, process, and maintain sensitive data. Accordingly, effective information security controls are essential to ensure that the agency's systems and information are protected from misuse and modification. Chapter 5 reviews the extent to which CDC has taken corrective actions to address the previously identified security program and technical control deficiencies and related recommendations for improvement.

Federal agencies are dependent on information systems to carry out operations. The risks to these systems are increasing as security threats evolve and become more sophisticated. To reduce the risk of a successful cyberattack, agencies can deploy intrusion detection and prevention capabilities on their networks and systems. Chapter 6 determined the reported effectiveness of agencies' implementation of the government's approach and strategy; the extent to which DHS and OMB have taken steps to facilitate the use of intrusion detection and prevention capabilities to secure federal systems; and the extent to which agencies reported implementing capabilities to detect and prevent intrusions.

Recent large-scale data breaches of public and private entities have put hundreds of millions of people at risk of identity theft or other harm. Chapter 7 reviews issues related to consumers' options to address risks of harm from data breaches and examines information and expert views on the effectiveness of consumer options to address data breach risks. While Chapter 8 considers the answer to this question: what legal obligations do Internet companies have to prevent and respond to data breaches? Then discusses several factors Congress might consider when weighing future legislation.

Chapter 1 - Federal agencies and our nation's critical infrastructures, such as communications and financial services, are dependent on information technology systems and electronic data to carry out operations and to process, maintain, and report essential information. The security of these systems and data is vital to public confidence and national security, prosperity, and well-being. Yet, cyber-based intrusions and attacks on federal and nonfederal systems have become not only more numerous and diverse, but also more damaging and disruptive. For example, a data breach reported in July 2015 at the Office of Personnel Management affected at least 21.5 million individuals and compromised federal employees' personal information, including Social Security numbers, residency and education history, employment history, financial history, and the fingerprints of approximately 5.6 million individuals. Due to cyber-based threats to federal systems and critical infrastructure, the persistent nature of information security vulnerabilities, and associated risks, the United States Government Accountability Office (GAO) have continued to designate information security as a government-wide high-risk area in our most recent biennial report to Congress—a designation GAO has made in each high-risk report since 1997. GAO expanded this area beyond federal information systems to include the protection of cyber critical infrastructure in 2003 and protecting the privacy of personally identifiable information in 2015. In December 2015, the President signed the Cybersecurity Information Sharing Act of 2015 (hereafter referred to as CISA or the act) into law to encourage the sharing of cyber threat information between the public and private sectors in a timely manner. The act designated seven federal agencies to coordinate and develop government-wide, publicly available policies, procedures, and guidance to assist federal and nonfederal entities in their efforts to receive and share cyber threat indicators and defensive measures. These seven agencies were the Departments of Homeland Security (DHS), Justice (DOJ), Defense (DOD), Commerce (DOC), Energy (DOE), and the Treasury, and the Office of the Director of National Intelligence (ODNI). In addition, the act included provisions related to ensuring the protection of privacy and civil liberties during the information sharing activities. It also incorporated language related to limitations on collection and use of personal information. According to the act, the designated agencies were to develop guidelines for the protection of privacy and civil liberties when implementing the CISA provisions. The act further required that these guidelines on privacy and civil liberties be consistent with the eight fair information practice principles defined in appendix A of the National Strategy for Trusted Identities in Cyberspace and other applicable provisions of law. The fair information practice principles are the widely accepted framework of defining principles to be used in the evaluation and consideration of systems, processes, or programs that affect individual privacy. Beyond the aforementioned requirements, the act included a provision for GAO to review actions taken by the federal government related to the removal of personal information from cyber threat indicators or defensive measures. Accordingly, our specific objective for this work was to determine the extent to which federal agencies have developed policies, procedures, and guidelines for the removal of personal information from cyber threat indicators and defensive measures, pursuant to CISA's provisions. To fulfill this objective, GAO gathered and analyzed policies, procedures, and guidelines that the seven designated federal agencies developed in response to requirements established within the act. GAO compared these documents to requirements established by the eight provisions in the act that relate to the removal of personal information from cyber threat indicators and defensive measures. GAO also compared the contents of these documents to the eight fair information practice principles. Further, GAO gathered and analyzed information from the seven designated federal agencies on their use of DHS's Automated Indicator Sharing (AIS) capability—the primary government-wide capability for real-time, automated receipt and sharing of cyber threat indicators and defensive measures that was called for in the act. Specifically, GAO looked at how each designated agency's relevant policies and guidance incorporated the use of this capability, including for the handling of personal information. In addition, GAO supplemented our analyses with interviews of responsible security and privacy officials from each of the seven agencies. GAO conducted this performance audit from February 2018 to October 2018 in accordance with generally accepted government auditing standards. Those standards require that GAO plan and perform the audit to obtain sufficient, appropriate evidence to provide a reasonable basis for our findings and conclusions based on our audit objectives. GAO believe that the evidence obtained provides a reasonable basis for our findings and conclusions based on our audit objectives.

Chapter 2 - The IRS has a demanding responsibility to collect taxes, process tax returns, and enforce the nation's tax laws. It relies extensively on computerized systems to support its financial and mission-related operations and on information security controls to protect the sensitive financial and taxpayer information that reside on those systems. As part of its audit of IRS's fiscal year 2017 and 2016 financial statements, GAO assessed whether controls over financial and tax processing systems were effective in ensuring the confidentiality, integrity, and availability of financial and sensitive taxpayer information. To do this, GAO examined IRS information security policies, plans, and procedures; tested controls over selected financial systems and applications; and interviewed key agency officials at four IRS locations.

Chapter 3 - IT systems are essential to the operations of the federal government. The supply chain—the set of organizations, people, activities, and resources that create and move a product from suppliers to end users—for IT systems is complex and global in scope. The exploitation of vulnerabilities in the IT supply chain is a continuing threat. Federal security guidelines provide for managing the risks to the supply chain. This testimony statement highlights information security risks associated with the supply chains used by federal agencies to procure IT systems. The statement also summarizes GAO's 2012 report that assessed the extent to which four national security-related agencies had addressed such risks. To develop this statement, GAO relied on its previous reports, as well as information provided by the national security-related agencies on their actions in response to GAO's previous recommendations. GAO also reviewed federal information security guidelines and directives.

Chapter 4 - The Office of Personnel Management (OPM) collects and maintains personal data on millions of individuals, including data related to security clearance investigations. In June 2015, OPM reported that an intrusion into its systems had affected the personnel records of about 4.2 million current and former federal employees. Then, in July 2015, the agency reported that a separate but related incident had compromised its systems and the files related to background investigations for 21.5 million individuals. From February 2015 through August 2017, United States Government Accountability Office (GAO) conducted multiple reviews of OPM's information security. GAO issued four reports based on these reviews. The reports contained 80 recommendations for improving the agency's security posture. The Explanatory Statement that accompanies the *Consolidated Appropriations Act, 2018*, included a provision for GAO to brief the House and Senate Appropriations Committees on actions taken by OPM in response to GAO's information security recommendations. Our specific objective was to determine the extent to which OPM has implemented our recommendations to improve the agency's information security. To accomplish this objective, GAO reviewed relevant documents and artifacts reflecting OPM's actions and progress toward implementing the 80 recommendations contained in the four reports, and assessed the actions against the intent of the recommendations. Our review examined the agency's actions taken on our recommendations through September 20, 2018. GAO also evaluated the agency's remedial action plans for those recommendations not yet implemented to determine the agency's intentions regarding the outstanding recommendations. Further, GAO interviewed officials in OPM's Office of the Chief Information Officer to obtain additional clarification about the agency's actions, as needed. On

September 26, 2018, GAO briefed members of your staff on the results of our review. This chapter formally transmits the final briefing slides. GAO conducted this performance audit from September 2018 through November 2018 in accordance with generally accepted government auditing standards. Those standards require that GAO plan and perform the audit to obtain sufficient, appropriate evidence to provide a reasonable basis for our findings and conclusions based on our audit objective. GAO believe that the evidence obtained provides a reasonable basis for our findings and conclusions based on our audit objective. In summary, OPM has made progress in implementing our recommendations for improving its security posture, but further actions are needed. As of September 20, 2018, the agency had implemented 51 (about 64 percent) of the 80 recommendations, but had not provided any evidence, or provided insufficient evidence, to demonstrate implementation of the remaining recommendations.

Chapter 5 - CDC is responsible for detecting and responding to emerging health threats and controlling dangerous substances. In carrying out its mission, CDC relies on information technology systems to receive, process, and maintain sensitive data. Accordingly, effective information security controls are essential to ensure that the agency's systems and information are protected from misuse and modification. GAO was asked to examine information security at CDC. In June 2018, GAO issued a limited official use only report on the extent to which CDC had effectively implemented technical controls and an information security program to protect the confidentiality, integrity, and availability of its information on selected information systems. This current report is a public version of the June 2018 report. In addition, for this public report, GAO determined the extent to which CDC has taken corrective actions to address the previously identified security program and technical control deficiencies and related recommendations for improvement. For this chapter, GAO reviewed supporting documents regarding CDC's actions on previously identified recommendations and interviewed personnel at CDC.

Chapter 6 - Federal agencies are dependent on information systems to carry out operations. The risks to these systems are increasing as security threats evolve and become more sophisticated. To reduce the risk of a successful cyberattack, agencies can deploy intrusion detection and prevention capabilities on their networks and systems. GAO first designated federal information security as a government-wide high-risk area in 1997. In 2015, GAO expanded this area to include protecting the privacy of personally identifiable information. Most recently, in September 2018, GAO updated the area to identify 10 critical actions that the federal government and other entities need to take to address major cybersecurity challenges. The federal approach and strategy for securing information systems is grounded in the provisions of the *Federal Information Security Modernization Act of 2014* and Executive Order 13800. The act requires agencies to develop, document, and implement an agency-wide program to secure their information systems. The Executive Order, issued in May 2017, directs agencies to use the National Institute of Standards and Technology's cybersecurity framework to manage cybersecurity risks. The *Federal Cybersecurity Enhancement Act of 2015* contained a provision for GAO to report on the effectiveness of the government's approach and strategy for securing its systems. GAO determined (1) the reported effectiveness of agencies' implementation of the government's approach and strategy; (2) the extent to which DHS and OMB have taken steps to facilitate the use of intrusion detection and prevention capabilities to secure federal systems; and (3) the extent to which agencies reported implementing capabilities to detect and prevent intrusions. To address these objectives, GAO analyzed OMB reports related to agencies' information security practices including OMB's annual report to Congress for fiscal year 2017. GAO also analyzed and summarized agency-reported security performance metrics and IG-reported information for the 23 civilian CFO Act agencies. In addition, GAO evaluated plans, reports, and other documents related to DHS intrusion detection and prevention programs, and interviewed OMB, DHS, and agency officials.

Chapter 7 - Recent large-scale data breaches of public and private entities have put hundreds of millions of people at risk of identity theft or other harm. GAO was asked to review issues related to consumers' options to address risks of harm from data breaches. This chapter, among other things, examines information and expert views on the effectiveness of consumer options to address data breach risks. GAO analyzed available data on options, collected and analyzed related documentation, conducted a literature review of studies, and interviewed a nongeneralizable sample of 35 experts (from academia, government entities, consumer and industry organizations) and identity theft service providers to reflect a range of views.

Chapter 8 - Recently, large Internet companies—i.e., companies that do most of their business on the Internet, such as social media platforms or search engines—have made headlines after failing to secure their users' personal information. For example, on September 28, 2018, Facebook announced a security breach affecting tens of millions of user accounts. According to Facebook, hackers exploited a vulnerability in its code that allowed them to steal "access tokens," which are the "equivalent of digital keys" that "keep people logged in to Facebook." Facebook later disclosed that, of the affected accounts, hackers accessed the names and contact details of 15 million users and the biographical information of another 14 million users. Just over a week after Facebook's breach, on October 8, 2018, Google, in announcing the end of its social network Google+, disclosed that a software glitch exposed the personal data associated with up to 500,000 Google+ accounts. Google explained that it discovered and resolved the glitch in March 2018 and that there was no evidence anyone misused the exposed data. The Internet search giant reportedly made an initial decision not to disclose the incident, before reversing course and shutting down the Google+ platform following a Wall Street Journal investigation. These incidents raise the question: what legal obligations do Internet companies have to prevent and respond to data breaches? This Sidebar considers the answer to this question under federal and state law. The Sidebar then discusses several factors Congress might consider when weighing future legislation.

In: Information Security
Editor: Paul Scholz
ISBN: 978-1-53615-880-9

Chapter 1

CYBERSECURITY: FEDERAL AGENCIES MET LEGISLATIVE REQUIREMENTS FOR PROTECTING PRIVACY WHEN SHARING THREAT INFORMATION*

United States Government Accountability Office

The Honorable Richard Burr
Chairman

The Honorable Mark R. Warner
Vice Chairman
Select Committee on Intelligence United States Senate

The Honorable Devin Nunes
Chairman

* This is an edited, reformatted and augmented Accessible version of the United States Government Accountability Office Report, Publication No. GAO-19-114R, dated December 6, 2018.

The Honorable Adam Schiff
Ranking Member
Permanent Select Committee on Intelligence
House of Representatives

Federal agencies and our nation's critical infrastructures, such as communications and financial services, are dependent on information technology systems and electronic data to carry out operations and to process, maintain, and report essential information.[1] The security of these systems and data is vital to public confidence and national security, prosperity, and well-being. Yet, cyber-based intrusions and attacks on federal and nonfederal systems have become not only more numerous and diverse, but also more damaging and disruptive. For example, a data breach reported in July 2015 at the Office of Personnel Management affected at least 21.5 million individuals and compromised federal employees' personal information, including Social Security numbers, residency and education history, employment history, financial history, and the fingerprints of approximately 5.6 million individuals.

Due to cyber-based threats to federal systems and critical infrastructure, the persistent nature of information security vulnerabilities, and associated risks, we have continued to designate information security as a government-wide high-risk area in our most recent biennial report to Congress—a designation we have made in each high-risk report since 1997.[2] We expanded this area beyond federal information systems to

---

[1] The term "critical infrastructure," as defined in the Critical Infrastructures Protection Act of 2001, part of the Uniting and Strengthening America by Providing Appropriate Tools Required to Intercept and Obstruct Terrorism Act of 2001 (USA PATRIOT Act), refers to systems and assets so vital to the United States that their incapacity or destruction would have a debilitating impact on security, national economic security, national public health or safety, or any combination of these matters. 42 U.S.C.§5195c(e). Federal policy identifies 16 critical infrastructure sectors: chemical; commercial facilities; communications; critical manufacturing; dams; defense industrial base; emergency services; energy; financial services; food and agriculture; government facilities; health care and public health; information technology; nuclear reactors, materials, and waste; transportation systems; and water and wastewater systems.

[2] GAO, *High-Risk Series: Progress on Many High-Risk Areas, While Substantial Efforts Needed on Others*, GAO-17- 317, (Washington, D.C.: February 15, 2017).

include the protection of cyber critical infrastructure in 2003 and protecting the privacy of personally identifiable information in 2015.

In December 2015, the President signed the Cybersecurity Information Sharing Act of 2015 (hereafter referred to as CISA or the act) into law to encourage the sharing of cyber threat information between the public and private sectors in a timely manner.[3] The act designated seven federal agencies to coordinate and develop government-wide, publicly available policies, procedures, and guidance to assist federal and nonfederal entities[4] in their efforts to receive and share[5] cyber threat indicators and defensive measures.[6] These seven agencies were the Departments of Homeland Security (DHS), Justice (DOJ), Defense (DOD), Commerce (DOC), Energy (DOE), and the Treasury, and the Office of the Director of National Intelligence (ODNI).

In addition, the act included provisions related to ensuring the protection of privacy and civil liberties during the information sharing activities. It also incorporated language related to limitations on collection and use of personal information. According to the act, the designated agencies were to develop guidelines for the protection of privacy and civil liberties when implementing the CISA provisions.

The act further required that these guidelines on privacy and civil liberties be consistent with the eight fair information practice principles defined in appendix A of the National Strategy for Trusted Identities in

---

[3] Consolidated Appropriations Act, 2016, Pub. L. No. 114-113, div. N (Cybersecurity Act of 2015), Title I (Cybersecurity Information Sharing Act of 2015), 129 Stat. 2242, 2936-56 (2015) (*codified at* 6 U.S.C. §§ 1501-10).

[4] A federal entity is a department or agency of the United States or any component of such department or agency. Nonfederal entities include state and local governments, private sector entities, and academic institutions.

[5] The act established requirements to develop artifacts for the receipt of cyber threat indicators and defensive measures by federal entities and for the sharing of cyber threat indicators and defensive measures by both federal and nonfederal entities. Unless explicitly specified as either receiving or sharing as part of a requirement within the act, this correspondence will collectively refer to these activities as "information sharing".

[6] As defined in CISA, cyber threat indicators include threat-related information such as methods of defeating or causing users to unwittingly enable the defeat of security controls and methods of exploiting cybersecurity vulnerabilities. Defensive measures include any actions, devices, procedures, techniques, or other means that detect, prevent, or mitigate a known or suspected cybersecurity threat or vulnerability.

Cyberspace and other applicable provisions of law.[7] The fair information practice principles are the widely accepted framework of defining principles to be used in the evaluation and consideration of systems, processes, or programs that affect individual privacy.

Beyond the aforementioned requirements, the act included a provision for GAO to review actions taken by the federal government related to the removal of personal information from cyber threat indicators or defensive measures. Accordingly, our specific objective for this work was to determine the extent to which federal agencies have developed policies, procedures, and guidelines for the removal of personal information from cyber threat indicators and defensive measures, pursuant to CISA's provisions.[8]

To fulfill this objective, we gathered and analyzed policies, procedures, and guidelines that the seven designated federal agencies developed in response to requirements established within the act. We compared these documents to requirements established by the eight provisions in the act (identified in table 1 of this chapter) that relate to the removal of personal information from cyber threat indicators and defensive measures. We also compared the contents of these documents to the eight fair information practice principles (identified in table 2 of this chapter).

Further, we gathered and analyzed information from the seven designated federal agencies on their use of DHS's Automated Indicator Sharing (AIS) capability—the primary government-wide capability for real-time, automated receipt and sharing of cyber threat indicators and defensive measures that was called for in the act.[9] Specifically, we looked

---

[7] The White House, *National Strategy for Trusted Identities in Cyberspace,* Appendix A (Washington, D.C.: April 2011).

[8] CISA refers to the removal of "personal information" from cyber threat indicators and defensive measures. For the purposes of this correspondence, we consider personal information to include all information that an individual would consider sensitive and would not want to be released publicly. This information is generally referred to as personally identifiable information (PII) and is similar to, but may not be identical to, personal information.

[9] The AIS initiative is an automated capability that receives, processes, and disseminates cyber threat indicators and defensive measures in real-time by enabling DHS to receive indicators from federal and nonfederal entities, remove personally identifiable information and other sensitive information not directly related to the cybersecurity threat, and disseminate the

at how each designated agency's relevant policies and guidance incorporated the use of this capability, including for the handling of personal information. In addition, we supplemented our analyses with interviews of responsible security and privacy officials from each of the seven agencies.

We conducted this performance audit from February 2018 to October 2018 in accordance with generally accepted government auditing standards. Those standards require that we plan and perform the audit to obtain sufficient, appropriate evidence to provide a reasonable basis for our findings and conclusions based on our audit objectives. We believe that the evidence obtained provides a reasonable basis for our findings and conclusions based on our audit objectives.

BACKGROUND

CISA encouraged the timely sharing of cyber threat information between the public and private sectors. To facilitate the implementation of provisions related to this information sharing, the act required the seven designated federal agencies to jointly develop publicly available policies, procedures, and guidelines to assist federal and nonfederal entities in sharing information on cyber threat indicators and defensive measures.[10]

Specifically, the act required DHS and DOJ to lead the development of procedures and guidelines to assist federal entities when receiving information on cyber threat indicators or defensive measures from federal and nonfederal entities. The act also required DHS and DOJ to work with DOD and ODNI to develop policies to facilitate the sharing of cyber threat indicators and defensive measures among federal and nonfederal entities. Further, the act required these four agencies to coordinate the development of these artifacts with DOC, DOE and the Treasury. In support of this

---

cyber threat indicators and defensive measures, as appropriate, to other federal and nonfederal entities.

[10] CISA designated these entities as the appropriate agencies to develop and coordinate regarding the required policies, procedures, and guidance.

information sharing, the White House released a memorandum in 2016 stating that all departments and agencies should use DHS's AIS capability for the purpose of sharing cyber threat information.

In addition, the act included specific provisions for the protection of privacy and civil liberties. In this regard, CISA required DHS and DOJ, in coordination with the other five designated federal agencies, to develop government-wide guidelines for the protection of privacy and civil liberties when implementing provisions of the act.

Among other things, the seven designated agencies were to:

- develop privacy and civil liberties guidance governing the receipt, retention, use, and dissemination of cyber threat indicators, consistent with the eight fair information practice principles;
- develop and implement the aforementioned real-time sharing capability in compliance with this privacy and civil liberties guidance; and
- incorporate appropriate security and privacy protections into all other policies, procedures, and guidance for receiving and sharing cyber threat indicators and defensive measures specified in the act.

Designated Federal Agencies Met Requirements to Develop Policies, Procedures, and Guidelines for Removal of Personal Information

The seven designated federal agencies developed policies, procedures, and guidelines that met the eight CISA provisions relevant to the removal of personal information from cyber threat indicators and defensive measures. Table 1 summarizes these CISA provisions and identifies the government-wide policies, procedures, and guidelines that the designated agencies created to fulfill each provision for the sharing of cyber threat indicators and defensive measures, and the removal of personal information.

Table 1. Government-wide Policies, Procedures, and Guidelines Created to Fulfill Eight Provisions of the Cybersecurity Information Sharing Act of 2015 Related to the Removal of Personal Information from Cyber Threat Indicators and Defensive Measures

CISA section[a]	Requirement	Related policy, procedure, or guidance created by designated agencies
103(a)	The Office of the Director of National Intelligence (ODNI), the Department of Defense (DOD), the Department of Justice (DOJ), and the Department of Homeland Security (DHS) are to develop and issue procedures for the timely sharing of cyber threat information to mitigate adverse effects	*Sharing of Cyber Threat Indicators and Defensive Measures by the Federal Government under the Cybersecurity Information Sharing Act of 2015,* February 16, 2016
103(b)	ODNI, DOD, DOJ, and DHS are to ensure that the secure real-time sharing of cyber threat information excludes personal information not directly related to a cybersecurity threat	*Sharing of Cyber Threat Indicators and Defensive Measures by the Federal Government under the Cybersecurity Information Sharing Act of 2015,* February 16, 2016
103(c)	Procedures developed shall be issued no later than 60 days after the date of enactment of CISA	*Sharing of Cyber Threat Indicators and Defensive Measures by the Federal Government under the Cybersecurity Information Sharing Act of 2015,* February 16, 2016
105(a)(1)	DOJ and DHS are to develop interim policies to guide in the receipt of cyber indicators and defensive measures by the federal government	*Interim Procedures Related to the Receipt of Cyber Threat Indicators and Defensive Measures by the Federal Government,* February 16, 2016
105(a)(2)	DOJ and DHS are to develop final policies to guide in the receipt of cyber indicators and defensive measures by the federal government	*Final Procedures Related to the Receipt of Cyber Threat Indicators and Defensive Measures by the Federal Government, June 16, 2016*
105(a)(4)	DOJ and DHS are to develop guidance to assist nonfederal entities and promote sharing of cyber threat indicators with federal entities	*Guidance to Assist Nonfederal Entities to Share Cyber Threat Indicators and Defensive Measures with Federal Entities under the Cybersecurity Information Sharing Act of 2015, February 16, 2016*

Table 1. (Continued)

CISA section[a]	Requirement	Related policy, procedure, or guidance created by designated agencies
105(b)(1)	DOJ and DHS are to develop interim guidelines relating to privacy and civil liberties that governs the receipt, retention, use, and dissemination of cyber threat information and are consistent with fair information practice principles	*Privacy and Civil Liberties Interim Guidelines: Cybersecurity Information Sharing Act of 2015, February 16, 2016*
105(b)(2)	DOJ and DHS are to develop final guidelines relating to privacy and civil liberties that governs the receipt, retention, use, and dissemination of cyber threat information and are consistent with fair information practice principles	*Privacy and Civil Liberties Final Guidelines: Cybersecurity Information Sharing Act of 2015, June 16, 2016b*

Source: GAO analysis of policies and procedures created by DOD, DOJ, DHS, and ODNI, and in response to CISA.

[a] CISA = Cybersecurity Information Sharing Act of 2015, and can be found at https://www.congress.gov/114/plaws/publ113/PLAW-114publ113.pdf.

[b] In accordance with CISA, DHS and DOJ produced an updated version of this document in June 2018. Changes between the two versions were non-substantial for the purposes of our analysis.

The table is followed by additional discussion of the provisions and related policies, procedures, and guidance. To satisfy subsections 103(a), (b), and (c) of CISA, which called for the development and issuance of procedures for the timely sharing of cyber threat information and defensive measures, DHS, DOJ, ODNI, and DOD, in coordination with the other three agencies designated in the act, developed the procedures in *Sharing of Cyber Threat Indicators and Defensive Measures by the Federal Government under the Cybersecurity Information Sharing Act of 2015*. As required by subsection 103(a), these procedures outline how federal entities are to share classified and unclassified cyber threat indicators and defensive measures in a timely manner and in a way that mitigates adverse effects from cybersecurity threats. For example, the procedures include details on existing government programs that facilitate the sharing of information on cybersecurity threats and the periodic publication of cybersecurity best practices—such as technical security guidelines for federal agency information systems, and recommended practices and references for technical and nontechnical users.

In addition, as required by subsection 103(b), the procedures include information on the existing roles and responsibilities of federal and nonfederal entities when sharing information. The procedures also incorporate requirements related to the security of cyber threat indicators and defensive measures prior to sending them to other entities, in order to assure that federal or nonfederal entities do not share personal information unnecessarily.[11] For example, the guidance includes requirements for notification when a cyber threat indicator or defensive measure is shared in error and for protection against unauthorized access to a cyber threat indicator or defensive measure. Further, as required by subsection 103(c), the designated agencies issued the procedures in less than 60 days after the enactment of the act.

DOJ and DHS met the requirements in sections 105(a)(1) and 105(a)(2) of CISA, which call for the development of policies for the receipt and sharing of cyber threat indicators and defensive measures, including consideration of privacy and civil liberties as a part of the receipt process. Specifically, in coordination with the other five designated federal agencies, these departments developed the interim and final versions of the guidance document, *Procedures Related to the Receipt of Cyber Threat Indicators and Defensive Measures by the Federal Government.* This document describes the processes for receiving, handling, and disseminating cyber threat indicators and defensive measures shared through DHS's AIS, as well as through the submission of a website form or email.[12] For all federal entities that receive cyber threat indicators and defensive measures, this guidance states and provides context for the statutory requirements for privacy and civil liberties, and for the maintenance of data on the number of cyber threat indicators and defensive measures where personal information concerns were addressed.

DOJ and DHS also addressed the provision in section 105(a)(4) through creation of the *Guidance to Assist Nonfederal Entities to Share Cyber Threat Indicators and Defensive Measures with Federal Entities*

---

[11] In some cases, personal information directly related to a cybersecurity threat may be shared.

[12] The final *Procedures Related to the Receipt of Cyber Threat Indicators and Defensive Measures by the Federal Government* was released in June 2016, replacing the February 2016 interim version.

under the CISA of 2015. Specifically, the stated intent of this guidance is to assist nonfederal entities in identifying cyber threat information. Further, the guidance explains how to share such information with federal entities through both DHS's AIS and other means authorized by CISA.

Table 2. Definitions of Fair Information Practice Principles

Principle	Definition
Transparency	Organizations should be transparent and notify individuals regarding the collection, use, dissemination, and maintenance of personally identifiable information (PII).
Individual participation	Organizations should involve the individual in the process of using PII and, to the extent practicable, seek individual consent for the collection, use, dissemination, and maintenance of PII. Organizations should also provide mechanisms for appropriate access, correction, and redress regarding the use of PII.
Purpose specification	Organizations should specifically articulate the authority that permits the collection of PII and specifically articulate the purpose or purposes for which the PII is intended to be used.
Data minimization	Organizations should only collect PII that is directly relevant and necessary to accomplish the specified purpose(s) and only retain PII for as long as is necessary to fulfill the specified purpose(s).
Use limitation	Organizations should use PII solely for the purpose(s) specified in the notice. Sharing PII should be for a purpose compatible with the purpose for which the PII was collected.
Data quality and integrity	Organizations should, to the extent practicable, ensure that PII is accurate, relevant, timely, and complete.
Accountability and auditing	Organizations should be accountable for complying with these principles, providing training to all employees and contractors who use PII, and auditing the actual use of PII to demonstrate compliance with these principles and all applicable privacy protection requirements against risks such as loss, unauthorized access or use, destruction, modification, or unintended or inappropriate disclosure.

Source: The White House, National Strategy for Trusted Identities in Cyberspace (Washington, D.C.: April 2011).

In response to requirements in sections 105(b)(1) and 105(b)(2) of CISA, on the development of guidelines for the consideration of privacy and civil liberties when sharing cyber threat indicators, DHS and DOJ, in coordination with the other five designated federal agencies, issued the

Privacy and Civil Liberties Interim Guidelines and the *Privacy and Civil Liberties Final Guidelines*.[13]

These guidance documents govern the receipt, retention, use, and dissemination of cyber threat information. For example, according to the *Final Guidelines*, federal entities should follow requirements to safeguard cyber threat indicators prior to sharing the indicators, including in cases when the shared cyber threat indicators contain specific individuals' personal information or information that identifies specific individuals that is not directly related to a cybersecurity threat.

Beyond the aforementioned provisions, CISA section 105(b) requires that the guidance documents created to fulfill its provisions also address the fair information practice principles as applicable. Table 2 defines the eight fair information practice principles.

The *Privacy and Civil Liberties Final Guidelines* fully meets requirements in section 105(b) by addressing all eight fair information practice principles. Specifically, the guidelines do so by establishing or considering the fair information practice principles as the primary guiding principles for all federal entity activities related to the receipt, retention, use, and dissemination of cyber threat indicators, as authorized by CISA. For example:

- *Transparency principle*. The final guidelines state that federal entities are to be transparent about their receipt, retention, use, and dissemination of cyber threat indicators under CISA.
- *Individual participation principle*. The final guidelines state that, given the nature of a cyber threat indicator, an individual whose personal information is directly related to a cybersecurity threat does not have the ability to consent to, or to be involved in, the process used to collect, access, or correct that information.

---

[13] In accordance with CISA, DHS produced an updated version of this document in June 2018. Changes between the two versions were non-substantial for the purposes of our analysis.

Each of the other guidance documents refers to the requirements in the Privacy and Civil Liberties Final Guidelines as a necessary part of their own requirements.

Finally, DHS's guidance includes other related information on the consideration of the fair information practice principles specifically in support of its implementation of AIS. Specifically, the Final Procedures Related to the Receipt of Cyber Threat Indicators and Defensive Measures provides several examples to guide the implementation of fair information practice principles when federal and nonfederal entities use AIS. For example:

- *Purpose specification principle.* The procedures provide guidance to federal entities on using an AIS profile to standardize cyber threat indicator information received while adhering to all relevant privacy and civil liberties requirements.
- *Data minimization principle.* The procedures require nonfederal entity submissions to conform to the defined AIS profile. This ensures that submissions include input fields most directly related to cyber threat indicators. Furthermore, the procedures guide entities to add or delete fields when considering changes to the AIS profile, in compliance with CISA's provisions that are designed to limit the receipt, retention, use, and dissemination of cyber threat indicators containing personal information.
- *Use limitation principle.* The procedures state that failure by federal employees to abide by usage requirements will result in individual sanctions such as loss of access to information and loss of a security clearance.

Agency Comments

We provided a draft of this chapter to DHS, DOJ, DOD, DOC, DOE, Treasury, and ODNI for review and comment. DHS, DOJ, and DOD provided technical comments, which we incorporated into the report as appropriate. The other four agencies stated via email that they had no comments on the draft report.

We are sending copies of this chapter to the Secretaries of Commerce, Defense, Energy, Homeland Security, and the Treasury; the Attorney General, the Director of National Intelligence, appropriate congressional committees, and other interested parties.

Sincerely yours,

Nick Marinos
Director, Cybersecurity & Data Protection Issues

In: Information Security
Editor: Paul Scholz

ISBN: 978-1-53615-880-9

Chapter 2

INFORMATION SECURITY: IRS NEEDS TO RECTIFY CONTROL DEFICIENCIES THAT LIMIT ITS EFFECTIVENESS IN PROTECTING SENSITIVE FINANCIAL AND TAXPAYER DATA*

United States Government Accountability Office

ABBREVIATIONS

CFO	chief financial officer
FIPS	Federal Information Processing Standard
FISCAM	Federal Information Systems Controls Audit Manual
FISMA	Federal Information Security Modernization Act
FMFIA	Federal Managers' Financial Integrity Act

* This is an edited, reformatted and augmented version of the United States Government Accountability Office Report to the Commissioner of Internal Revenue, Publication No. GAO-18-391, dated July 2018.

ID	identification
IRS	Internal Revenue Service
NIST	National Institute of Standards and Technology
OMB	Office of Management and Budget
PIV	personal identity verification
SOC	service organization controls
Treasury	Department of the Treasury

Why GAO Did This Study

The IRS has a demanding responsibility to collect taxes, process tax returns, and enforce the nation's tax laws. It relies extensively on computerized systems to support its financial and mission-related operations and on information security controls to protect the sensitive financial and taxpayer information that reside on those systems.

As part of its audit of IRS's fiscal year 2017 and 2016 financial statements, GAO assessed whether controls over financial and tax processing systems were effective in ensuring the confidentiality, integrity, and availability of financial and sensitive taxpayer information. To do this, GAO examined IRS information security policies, plans, and procedures; tested controls over selected financial systems and applications; and interviewed key agency officials at four IRS locations.

What GAO Recommends

In addition to the prior recommendations that have not been implemented, GAO is recommending that IRS take 5 additional actions to more effectively implement security-related policies and plans. In a separate report with limited distribution, GAO is recommending 32 actions that IRS can take to address newly identified control deficiencies. In commenting on a draft of this chapter, IRS agreed with GAO's recommendations and stated that it would review each of the recommendations and ensure that its corrective actions include a root cause analysis for sustainable fixes that implement appropriate security controls.

WHAT GAO FOUND

The Internal Revenue Service (IRS) has made progress in resolving a number of previously reported control deficiencies. During fiscal year 2017, the agency made improvements in access controls by, for example, restricting unnecessary user access to certain applications and enforcing strong encryption on certain systems. IRS also corrected a previously identified contingency planning weakness for one system.

Nevertheless, continuing and newly identified control deficiencies limited the effectiveness of security controls for protecting the confidentiality, integrity, and availability of IRS's financial and tax processing systems. For example, IRS did not consistently (1) implement access controls by enforcing password expirations and minimum password lengths or by updating expiration dates for contractor passwords; (2) apply configuration management controls by documenting authorizations and approvals for changes to mainframe data and processing, or by installing critical security patches on multiple devices; and (3) implement certain components of its security program by correcting weaknesses in procedures or by updating system security plans. GAO has made recommendations to IRS to correct the identified security control deficiencies (see table). However, many deficiencies have not been corrected, and a large number of recommendations remained open at the conclusion of the audit of IRS's financial statements for fiscal year 2017.

Until IRS takes additional steps to address unresolved and newly identified control deficiencies and effectively implements components of its information security program, IRS financial reporting and taxpayer data will remain unnecessarily vulnerable to inappropriate and undetected use, modification, or disclosure. These shortcomings were the basis for GAO's determination that IRS had a significant deficiency in internal control over financial reporting systems for fiscal year 2017.

Status of GAO Information Security Control Recommendations to IRS to Correct Control Deficiencies at the Conclusion of Fiscal Year 2017

Information security control area	Prior recommendations open at the beginning of FY 2017	Prior recommendations closed at the end of FY 2017	New recommendations resulting from FY 2017audit	Total outstanding recommendations at the end of FY 2017
Access controls	120	(35)	21	106
Configuration management	29	(10)	13	32
Segregation of duties	1	(0)	0	1
Contingency planning	2	(1)	1	2
Security program	14	(3)	2	13
Total	166	(49)	37	154

Legend: FY = fiscal year.

Source: GAO analysis of Internal Revenue Service (IRS) data. | GAO-18-391.

July 31, 2018

The Honorable David J. Kautter
Acting Commissioner of Internal Revenue

The Internal Revenue Service (IRS) has a demanding responsibility to collect taxes, process tax returns, and enforce the nation's tax laws. A bureau of the U.S. Department of the Treasury, IRS relies extensively on computer systems to support its financial and mission-related operations. IRS also relies on information system security controls to protect the confidentiality, integrity, and availability of the sensitive financial and taxpayer information that resides on its systems.

GAO audits IRS's financial statements in accordance with authority conferred by the *Chief Financial Officers (CFO) Act of 1990*, as amended by the *Government Management Reform Act of 1994*. As part of our audit of IRS's fiscal year 2017 and 2016 financial statements, we assessed the effectiveness of the agency's information system security controls over selected financial and tax processing systems, information, and interconnected networks at four IRS locations. These systems support the processing, storage, and transmission of sensitive financial and taxpayer data.

On November 9, 2017, we issued our report on the results of our audit of IRS's fiscal year 2017 and 2016 financial statements, and on the effectiveness of its internal control over financial reporting[1] as of September 30, 2017.[2] During that audit, we identified continuing and new internal control deficiencies[3] concerning IRS's financial reporting systems that are important enough to merit the attention of those charged with governance of IRS. Therefore, we considered these continuing and new issues affecting IRS's internal control over financial reporting systems collectively to be a significant deficiency[4] in internal control in fiscal year 2017.[5]

Our objective for that audit was to assess whether IRS's controls over its financial and tax processing systems were effective in ensuring the confidentiality, integrity, and availability of sensitive financial and taxpayer data. To accomplish the objective, we examined the agency's information security policies, plans, and procedures;[6] tested controls over selected financial systems; reviewed previously reported control deficiencies; and assessed the effectiveness of corrective actions taken. We also interviewed key agency officials responsible for managing and

---

[1] We audited IRS's internal control over financial reporting as of September 30, 2017, based on criteria established under 31 U.S.C. § 3512(c), (d), commonly known as the *Federal Managers' Financial Integrity Act* (FMFIA).

[2] GAO, *Financial Audit: IRS's Fiscal Years 2017 and 2016 Financial Statements*, GAO-18-165 (Washington, D.C.: Nov. 9, 2017).

[3] A deficiency in internal control exists when the design or operation of a control does not allow management or employees, in the normal course of performing their assigned functions, to prevent, or detect and correct, misstatements on a timely basis.

[4] A significant deficiency is a deficiency, or a combination of deficiencies, in internal control that is less severe than a material weakness, yet important enough to merit the attention of those charged with governance. A material weakness is a deficiency, or combination of deficiencies, in internal control over financial reporting, such that there is a reasonable possibility that a material misstatement of the entity's financial statements will not be prevented, or detected and corrected, on a timely basis.

[5] GAO-18-165.

[6] Most of IRS's information security policies, plans, and procedures are non-public documents that cannot be described in detail in publicly available audit reports.

operating the selected systems. The focus of our evaluation was limited to systems relevant to financial management and reporting. Appendix I provides additional details on our objective, scope, and methodology.

We performed our work in accordance with generally accepted government auditing standards. Those standards require that we plan and perform the audit to obtain sufficient appropriate evidence to provide a reasonable basis for our findings and conclusions based on our audit objectives. We believe that the evidence obtained provides a reasonable basis for our findings and conclusions based on our audit objective.

This chapter is an integral part of our audit of IRS's fiscal years 2017 and 2016 financial systems. In this regard, the report presents the details of information system security control deficiencies we identified as part of our fiscal year 2017 audit of IRS's financial statements, and recommendations for corrective actions to address them. Specifically, the report highlights some of our new internal control deficiencies that we identified during our fiscal year 2017 testing of information system security controls over financial systems that are relevant to IRS's internal control over financial reporting.[7]

This chapter also presents details of one information system security control deficiency we identified during our audit that does not directly affect IRS's financial reporting, but does merit management's attention due to its potential effect on information system security controls for ensuring the confidentiality, integrity, and availability of sensitive information.[8] In

---

[7] An entity's internal control over financial reporting is a process effected by those charged with governance, management, and other personnel, the objectives of which are to provide reasonable assurance that (1) transactions are properly recorded, processed, and summarized to permit the preparation of financial statements in accordance with U.S. generally accepted accounting principles, and assets are safeguarded against loss from unauthorized acquisition, use, or disposition, and (2) transactions are executed in accordance with provisions of applicable laws, including those governing the use of budget authority, regulations, contracts, and grant agreements, noncompliance with which could have a material effect on the financial statements.

[8] Information system security controls include logical and physical access controls, configuration management, segregation of duties, continuity of operations, and security management. These controls are designed to ensure that access to data is appropriately restricted, physical access to sensitive computing resources and facilities is restricted, systems are securely configured to avoid exposure to known vulnerabilities, and incompatible duties are segregated among individuals. In addition, controls should ensure that backup and recovery plans are adequate and tested to ensure the continuity of essential operations, and that

addition, this chapter includes the results of our follow-up on the status of the agency's corrective actions to address information system security control deficiencies and associated recommendations contained in our prior years' reports that remained open at the beginning of our fiscal year 2017 audit.

The purpose of this chapter is solely to describe the scope and results of our testing of the effectiveness of IRS's security controls over its financial and tax processing systems relevant to financial management and reporting. Accordingly, this chapter is not suitable for any other purpose.

BACKGROUND

As the tax collector of the United States, the IRS mission is to help taxpayers understand and meet their tax responsibilities and to enforce tax laws with integrity and fairness. According to publicly available IRS data, in fiscal year 2017, the agency collected about $3.4 trillion in federal tax payments, processed a total of about 201 million returns, and paid about $437 billion in refunds and outlays. IRS employs over 81,000 year-round and seasonal staff in its Washington, D.C. headquarters, in offices in every state and U.S. territory, and in a few U.S. embassies and consulates. The agency also operates enterprise computing centers in Martinsburg, West Virginia and Memphis, Tennessee.

In carrying out its mission and responsibilities of administering tax laws, IRS relies extensively on computer systems to support its financial and mission-related operations. As such, it must ensure that its computer systems are effectively secured to protect sensitive financial and taxpayer data gathered when the agency is collecting taxes, processing tax returns, and enforcing tax laws.

IRS also collects and maintains a significant amount of personal and financial information on each U.S. taxpayer. Protecting this sensitive

---

security is managed entity-wide under a framework that provides a continuing cycle of activity for managing risk, developing security policies, assigning responsibilities, and monitoring the adequacy of the entity's computer-related controls.

information is essential to protecting taxpayers' privacy and preventing financial loss and damages that could result from identity theft and other financial crimes. IRS has an important responsibility for protecting sensitive information and implementing effective information system security controls, a component of internal control over financial reporting.

Without effective security controls, computer systems are vulnerable to human actions committed in error or with malicious intent. People acting with malicious intent can use their access to obtain sensitive information, commit fraud and identity theft, disrupt operations, or launch attacks against other computer systems and networks. These threats to computer systems and related critical infrastructure can come from sources that are internal and external to an organization. Internal threats include equipment failure, human errors, and fraudulent or malevolent acts by employees or contractors. External threats include the ever-growing number of cyber-based attacks that can come from a variety of sources such as individuals, groups, and countries who wish to do harm to an organization's systems.

Our previous reports, and those by federal inspectors general, describe persistent information security weaknesses that place federal agencies, including IRS, at risk of disruption, fraud, or inappropriate disclosure of sensitive information. In October 2017, the Treasury Inspector General for Tax Administration stated that security over taxpayer data and protection of IRS resources was the top priority in its list of top management challenges for the agency for fiscal year 2018.[9] Furthermore, since 1997, we have designated federal information security as a government-wide high-risk area.[10]

---

[9] Department of the Treasury, Treasury Inspector General for Tax Administration, *Management and Performance Challenges Facing the Internal Revenue Service for Fiscal Year 2018*, Memorandum for Secretary Mnuchin (Washington, D.C.: October 2017).

[10] GAO, *High-Risk Series: Information Management and Technology*, GAO/HR-97-9 (Washington, D.C.: February 1997) and *High-Risk Series: Progress on Many High-Risk Areas, While Substantial Efforts Needed on Others*, GAO-17-317 (Washington, D.C.: February 2017).

Federal Law and Guidance Provide a Framework for Protecting Federal Information and Systems

Information security programs and practices performed by an agency are essential to creating and maintaining effective internal controls within an organization's critical information technology infrastructure. The *Federal Managers' Financial Integrity Act*[11] requires the Comptroller General to issue standards for internal control in the federal government. These standards provide the overall framework for establishing and maintaining an effective internal control system and describe internal control as a process put in place by an entity's oversight body, management, and other personnel that provides reasonable assurance that the objectives (operations, reporting, and compliance) of an entity will be achieved.[12]

Information system security controls consist of those internal controls that are dependent on information systems processing, and include general controls (such as managing security, appropriately restricting access to data and systems, securely configuring systems, segregating incompatible duties, and planning for continuity of operations) at the entity-wide, system, and business process application levels; business process application controls (input, processing, output, interface, and data management system controls); and user controls (controls performed by people interfacing with information systems).

Federal law and guidance specify requirements for protecting federal information and systems. The *Federal Information Security Modernization Act of 2014* (FISMA)[13] is intended to provide a comprehensive framework

---

[11] Pub. L. No. 97-255, 96 Stat. 814 (1982). The *Federal Managers' Financial Integrity Act* (FMFIA) was codified at 31 U.S.C. § 3512.

[12] GAO, *Standards for Internal Control in the Federal Government*, GAO-14-704G (Washington, D.C.: September 2014).

[13] The Federal Information Security Modernization Act of 2014 (FISMA 2014) (Pub. L. No. 113-283, Dec. 18, 2014) largely superseded the Federal Information Security Management Act of 2002 (FISMA 2002), enacted as Title III, E-Government Act of 2002, Pub. L. No. 107-347, 116 Stat. 2899, 2946 (Dec. 17, 2002). As used in this report, FISMA refers to the new requirements in FISMA 2014, and to other relevant FISMA 2002 requirements that were unchanged by FISMA 2014 and continue in full force and effect.

for ensuring the effectiveness of information system security controls over information resources that support federal operations and assets. To accomplish this, FISMA requires each agency to develop, document, and implement an agency-wide information security program to provide security for the information and systems that support the operations and assets of the agency, using a risk-based approach.

Such a program includes assessing risk; developing and implementing cost-effective security controls, policies, and procedures; providing security awareness training; testing and evaluating the effectiveness of controls; planning, implementing, evaluating, and documenting remedial actions to address information security deficiencies; implementing procedures for detecting, reporting, and responding to security incidents; and ensuring continuity of operations. Beyond establishing these information security program requirements, the act also assigned the National Institute of Standards and Technology (NIST) the responsibility for developing standards and guidelines that include minimum information security requirements.

IRS Made Progress in Addressing Previously Reported Control Deficiencies, but Sensitive Financial and Taxpayer Data Continue to Be at Risk

IRS has addressed numerous information system security control issues in response to previously reported control deficiencies and our related recommendations. However, continuing and newly identified control deficiencies limited the effectiveness of the agency's implementation of access controls, configuration management, segregation of duties, contingency planning, and components of its information security program, to prevent or timely detect material misstatements and protect the confidentiality, integrity, and availability of its financial and tax processing systems and information.

Table 1. Status of GAO Information System Security Control Recommendations to IRS to Correct Control Deficiencies at the Conclusion of Fiscal Year 2017

Information security control area	Prior recommendations not implemented at the beginning of fiscal year 2017	Prior recommendations implemented or considered no longer relevant at the end of fiscal year 2017[a]	Prior recommendations not fully implemented at the end of fiscal year 2017	New recommen-dations resulting from fiscal year 2017 audit	Total outstanding recommendations at the end of fiscal year 2017
Access controls					
Boundary protection	11	(0)	11	—	11
Identification and authentication	35	(9)	26	13	39
Authorization	22	(6)	16	3	19
Cryptography	34	(13)	21	—	21
Audit and monitoring	12	(5)	7	5	12
Physical security	6	(2)	4	—	4
Total access controls	120	(35)	85	21	106
Configuration management	29	(10)	19	13	32
Segregation of duties	1	(0)	1	—	1
Contingency planning	2	(1)	1	1	2
Information security program					
Risk assessments	1	(0)	1	—	1
Policies and procedures	5	(1)	4	—	4
Security plans	1	(0)	1	2	3
Training	1	(1)	0	—	0

Information security control area	Prior recommendations not implemented at the beginning of fiscal year 2017	Prior recommendations implemented or considered no longer relevant at the end of fiscal year 2017[a]	Prior recommendations not fully implemented at the end of fiscal year 2017	New recommen-dations resulting from fiscal year 2017 audit	Total outstanding recommendations at the end of fiscal year 2017
Testing and evaluation	5	(1)	4	—	4
Remedial actions	1	(0)	1	—	1
Total information security program	14	(3)	11	2	13
Grand total	166	(49)	117	37	154

Legend: — = no recommendations made.

Source: GAO analysis of Internal Revenue Service (IRS) data. | GAO-18-391.

[a]We did not consider certain control deficiencies to be corrected or mitigated; rather the issues were no longer relevant due to IRS's changing operating environment.

As indicated in table 1, we have previously made a number of recommendations to IRS to correct these control deficiencies, some of which still remain outstanding, in addition to making new recommendations based on our audit of IRS's fiscal year 2017 financial statements.

IRS Improved Access Controls, but Deficiencies Remained

A basic management objective for any organization is to protect the resources that support its critical operations from unauthorized access. This is accomplished by designing and implementing controls to prevent, limit, and detect unauthorized access to programs, data, facilities, and other computing resources. Access controls include both logical and physical controls related to the (1) protection of system boundaries, (2) identification and authentication of users, (3) authorization of access permissions, (4) encryption of sensitive information, (5) audit and monitoring of system activity, and (6) physical security of facilities and computing resources.

Deficiencies in IRS's Network Boundary Protection Continued to Exist

Boundary protection controls the logical connectivity into and out of networks and to and from devices attached to the network. Unnecessary connectivity to an organization's network increases not only the number of access paths that must be managed and the complexity of the task, but also the risk of unauthorized access in a shared environment.

IRS had developed and documented policies for protecting system boundaries. The *Internal Revenue Manual* requires that communications be monitored and controlled at the external boundary and at key internal boundaries within its systems. The manual also requires that management traffic transmitted across an Internet protocol backbone network be encrypted to ensure confidentiality and integrity. In addition, NIST recommends that devices be identified and authenticated prior to establishing a connection, and that approved authorizations for controlling the flow of information between interconnected systems should be enforced.[1]

However, IRS did not correct our previously reported boundary control deficiencies, such as not implementing access control lists on certain network devices to prevent unauthorized users from logging into the network devices; and not ensuring that authenticated network protocols were being used on its network devices. Until IRS corrects these deficiencies to its network boundaries, increased risk exists that its network devices and systems could be compromised, which could affect system availability.

IRS Inconsistently Implemented Identification and Authentication Controls for Financial Systems

Identification is the process of distinguishing one user from others as a prerequisite for granting access to resources in an information system. User identification (ID) is important because it is the means by which specific access privileges are assigned and recognized by the computer. However, the confidentiality of a user ID is typically not protected. For this reason, other means of authenticating users—that is, determining whether individuals are who they claim to be—are typically implemented.

Organizations may use other means of authenticating users to determine whether individuals are who they claim to be, such as tokens or biometrics.

IRS had developed and documented policies for identification and authentication. Specifically, the *Internal Revenue Manual* requires password complexity for all IRS information systems with password-based authentication. The manual also requires passwords that are not found in the dictionary and contain at least one numeric character, one special character, one uppercase letter, and one lowercase letter. Passwords for people must be set to expire within 90 days and passwords for service

---

[1] National Institute of Standards and Technology, *Security and Privacy Controls for Federal Information Systems and Organizations*, Special Publication 800-53, Revision 4 (Gaithersburg, Md.: April 2013).

accounts must be set to expire within 366 days. In addition, the manual requires that service account passwords have a minimum length of 14 characters. Further, the manual requires the use of a certificate revocation list to identify personal identity verification (PIV) certificates[2] that have been revoked.

IRS improved identification and authentication by enforcing password complexity for several user- and system-level accounts on various servers and by setting password expiration parameters for user and service accounts on several servers and databases. Nevertheless, deficiencies in authentication persisted. For example, IRS did not:

- enforce password expiration limits for several applications we reviewed;
- enforce minimum password lengths for service accounts supporting several applications we reviewed; and
- enable certificate revocation lists to check PIV certificates for user authentication to a financial system.

Until IRS fully remediates authentication control deficiencies, it is at increased risk that controls could be compromised, permitting unauthorized access to its systems and data.

IRS Did Not Always Limit Authorization of User Access Rights and Privileges to Only Personnel Who Required It to Perform Their Jobs

Authorization is the process of granting access rights and privileges to a system or a file. Access rights and privileges specify what a user can do after being authenticated to the information system, allowing the authorized user to read or write to files and directories. A key component

[2] Personal identity verification (PIV) card is a physical identity card, such as a "smart" card, issued to an individual that contains stored identity credentials, such as a photograph, cryptographic keys, or digitized fingerprint, used to verify the identity of the cardholder against the stored credentials by another person or an automated process. PIV certificates can be used for authentication to verify that PIV credentials were issued by an authorized entity, had not expired, and had not been revoked, and that the holder of the credentials was the same individual to whom the PIV card was issued.

of authorization is the concept of "least privilege," which means that users should be granted the least amount of privileges necessary to perform their duties. Maintaining access rights and privileges is one of the most important aspects of administering systems security.

IRS had developed and documented policies for authorizing access to its systems. According to the *Internal Revenue Manual*, system access is to be granted based on the principle of least privilege—that is, the minimum access necessary to perform one's duties. The manual also requires that user accounts be reviewed for compliance annually, that passwords for contractors expire at the end of the contract period of performance minus one day, and that authorizations for accounts, including non-unique accounts be approved and maintained.

IRS corrected 4 of the 22 authorization control deficiencies that we previously identified. As an example, the agency restricted unnecessary user access on Oracle databases supporting an application. The agency also restricted excessive user privileges to another application by limiting users' ability to enter certain database commands.

However, authorization control deficiencies still existed in IRS's computing environment. For example, IRS entered incorrect expiration dates for contractor passwords in the system used for managing user access authorizations. Specifically, the agency set passwords for 10 contractor profiles in production and 197 contractor profiles in the test environment of the mainframe to expire on dates that extended beyond the end of the contract period of performance.

In addition, IRS did not maintain and approve authorizations for 20 non-unique accounts that were used for its training environment.[3] Specifically, the agency did not provide documentation that authorizations for any of the 20 accounts had been approved or reviewed annually.

These weaknesses place the agency at increased risk that users with excessive privileges, users who should no longer have access to a system, and unauthorized users could inadvertently or deliberately access and

---

[3] A "non-unique" account is a user account that does not have a specific person assigned to that account.

modify systems. These risks jeopardize the confidentiality and integrity of the data they contain.

IRS Made Limited Progress in Correcting Control Deficiencies for Encryption of Sensitive Information

Cryptography can be used in identification and authorization to protect the integrity and confidentiality of computer programs and data in transmission or storage. Using algorithms (mathematical functions) and keys (strings of seemingly random bits), cryptographic modules[4] (1) encrypt a message or file so that it is unintelligible to those who do not have the secret key needed to decrypt it, thus keeping the contents of the message or file confidential; (2) provide an electronic signature that can be used to determine if any changes have been made to the related file, thus, ensuring the file's integrity; or (3) link a message or document to a specific individual's or group's key, thus, ensuring that the "signer" of the file can be identified.

IRS had developed and documented policies for encrypting data. Specifically, the *Internal Revenue Manual* requires that the agency use cryptographic mechanisms to prevent the unauthorized disclosure of information (confidentiality) and to detect changes to information (integrity). The manual also requires that IRS implement encryption mechanisms for authentication to a cryptographic module that meet the requirements of applicable federal laws, executive orders, directives, policies, regulations, standards, and guidance for such authentication. One such standard is the Federal Information Processing Standard (FIPS) 140-2,[5] which is used for encryption.

IRS corrected 13 of the 34 encryption control deficiencies that we previously identified. For example, the agency encrypted the authentication from its workstations to one of its systems. In addition, the agency configured various platforms and client software to encrypt connections

---

[4] A cryptographic module is the set of hardware, software, firmware, or some combination thereof that implements cryptographic logic or processes, including algorithms, and is contained within the encrypted boundary of the module.

[5] National Institute of Standards and Technology, *Security Requirements for Cryptographic Modules*, Federal Information Processing Standard 140-2, (Gaithersburg, Md.: May 2001).

between systems, and used encryption on servers supporting several systems.

However, IRS had not yet addressed 21 of 34 recommendations that we previously made to the agency concerning encryption control deficiencies. For example, it had not enforced the use of FIPS 140-2 compliant encryption algorithms on some systems and applications. In addition, it had not yet encrypted sensitive data on its Oracle databases supporting 11 systems and applications we previously reviewed. As a result, the agency has an increased risk that an unauthorized individual could exploit encryption weaknesses to view and then use data, such as user IDs and passwords, to gain access to systems that contain financial and sensitive data.

IRS Has Made Limited Progress Enhancing the Audit and Monitoring Controls of Its Financial Systems

Audit and monitoring involves the regular collection, review, and analysis of auditable events for indications of inappropriate or unusual activity and the appropriate investigation and reporting of such activity. Automated mechanisms may be used to integrate audit monitoring, analysis, and reporting into an overall process for investigation and response to suspicious activities. Audit and monitoring controls can help information systems security professionals routinely assess computer security, recognize an ongoing attack, and perform investigations during and after an attack.

IRS had developed and documented policies for auditing and monitoring its systems, and had issued an interim update to its *Internal Revenue Manual*.[6] The manual requires that IRS information systems generate audit records containing details to facilitate the reconstruction of events if unauthorized activity occurs. In addition, it requires the agency to review and analyze these audit records to determine if inappropriate or unusual activity has occurred. Further, the agency's procedures for audit

---

[6] IRS issued an interim guidance memorandum rescinding Internal Revenue Manual 10.8.3, which was the agency's previous standard for auditing and monitoring systems. The interim guidance made Internal Revenue Manual 10.8.1 the authority for these requirements.

log event identification, analysis, and reporting, require audit plans to be developed, approved, and implemented. Finally, the *Internal Revenue Manual* requires reviews of user and service accounts for compliance with account management requirements to ensure that accounts are still necessary and configured properly.

However, IRS has made limited progress in enhancing its audit and monitoring capabilities. For example, the agency corrected three previously identified weaknesses by reconfiguring the audit trails for several of its databases supporting three applications to enable the reconstruction of specific actions. Nevertheless, deficiencies in the agency's audit and monitoring capabilities persist. In this regard, the agency had not fully implemented 7 of 12 recommendations we previously made to correct deficiencies identified in audit and monitoring controls.[7] For example, the agency had neither updated nor implemented its audit and monitoring requirements in several system and application audit plans. Also, it had not enabled database logging, nor reviewed, analyzed, or reported auditable and actionable events on a database supporting a tax payment system. In addition, IRS did not consistently detect improperly configured encryption settings for user and service accounts or detect configuration changes made to the mainframe.

Without effective audit and monitoring controls, IRS's ability to establish individual accountability, monitor compliance with security and configuration management policies, and identify anomalous activity is reduced.

IRS Improved Physical Security Controls, but Prior Deficiencies Remain

Physical security controls are important for protecting computer facilities and resources from espionage, sabotage, damage, and theft.

---

[7] As of September 30, 2017, we determined that IRS had addressed 4 of the 12 recommendations. For 7 of the remaining 8 recommendations, IRS had not completed corrective actions. As for the remaining recommendation, we determined that it was no longer relevant due to the changing operating environment and issued a new specific recommendation that more accurately reflects addressing the associated deficiency in the current environment.

They include, among other things, policies and practices for authorizing individuals' physical access to facilities and resources; and periodically reviewing access authorizations in order to ensure that continued access is necessary. At IRS, control measures, such as access cards that are used to permit or deny access to certain areas of a facility, are vital to safeguarding its computing resources.

IRS had developed and documented policies for physically protecting its computer resources. In this regard, the *Internal Revenue Manual* requires access controls to protect employees and contractors, information systems, and the facilities in which the personnel and systems are located. The manual also requires that department managers of restricted areas approve all names added to the authorized access list for restricted areas; and that they review, validate, sign, and date the list monthly and then forward the list to the physical security office for review.

IRS had implemented multiple physical security controls at its enterprise computing centers to safeguard assets against possible theft and malicious actions. For example, the agency implemented security measures to control physical access to restricted areas at its computing center with the use of badge sensors and keypads for card/pin number credentialing. It also corrected a previously identified deficiency by ensuring that network equipment in restricted areas was housed in locked cabinets.

However, IRS had not corrected previously identified deficiencies regarding effectively reviewing access lists of individuals with an ongoing need to access restricted areas at two computing centers. The access lists continued to include individuals who no longer required access, and who should have been removed from the lists. Because individuals may be allowed inappropriate access to restricted areas, IRS has reduced assurance that its computing resources and sensitive information are protected from unauthorized access.

IRS Improved Configuration Management Controls, but Deficiencies Remained

Configuration management administers security features for all hardware, software, and firmware components of an information system throughout its life cycle. Effective configuration management provides reasonable assurance that systems are operating securely and as intended. It encompasses policies, plans, and procedures that call for proper authorization, testing, approval, and tracking of all configuration changes; and for timely software updates to protect against known vulnerabilities. Ineffective configuration management controls increase the risk that unauthorized changes could occur and that systems are not protected against known vulnerabilities.

IRS Had Not Documented Authorizations and Approvals of Changes To Mainframe Data and Processing

IRS had developed and documented policies for managing changes to its systems. Specifically, the *Internal Revenue Manual* requires that all changes to production systems and processing be authorized in advance and that approvals be documented.

Nevertheless, IRS was unable to provide supporting documentation for 13 changes made to critical mainframe datasets. In addition, as in previous years, the agency continued to alter production data processing on the mainframe—including tax data processing—outside established change control procedures. The lack of effective change management increases the agency's risk that unauthorized changes can be made to applications that result in the loss of data or program integrity.

IRS Had Not Applied Critical Security Patches and Used Unsupported Software on Multiple Devices

IRS also had developed and documented policies for managing the configuration of its information technology systems. The *Internal Revenue Manual* requires that the agency manage systems to reduce vulnerabilities by, among other things, installing patches in accordance with the timelines defined in its policy, based on the criticality of the updates and patches. The manual also requires that database software be removed or updated before the vendor discontinues support.

During 2017, the agency had not installed critical patch updates to a recently upgraded database supporting an important IRS information system, nor had it addressed deficiencies related to installing critical patch updates identified in prior years. Specifically, the agency still had not applied critical security patches to databases supporting five information systems, including its personnel and payroll system, or to servers supporting eight information systems, including its general ledger system.

In addition, IRS continued to rely on database software that was no longer supported by the vendor. Such reliance is problematic because vendors generally do not provide updates for unsupported software even if vulnerabilities are known. By not installing patches and replacing unsupported software per its own requirements, IRS has increased the risk that individuals may exploit known vulnerabilities in its systems.

Segregation of Duties and Contingency Planning Control Deficiencies Remained

In addition to access controls and configuration management, other controls should be in place to ensure the confidentiality, integrity, and availability of an organization's information. These controls include policies, procedures, and techniques for segregating incompatible duties and planning for continuity of operations.

IRS Had Not Corrected a Prior Segregation of Duties Deficiency for One Financial System

Segregation of duties helps to ensure that no single individual can independently control all key aspects of a process or computer-related operation and, thereby, make unauthorized changes. Segregation of duties increases the likelihood that errors and wrongful acts will be detected, because the activities of one individual or group will serve as a check on the activities of the other. Conversely, inadequate segregation of duties increases the risk that erroneous or fraudulent transactions could be processed, improper program changes implemented, and computer resources damaged or destroyed.

IRS had developed and documented policies for dividing and separating incompatible duties and responsibilities. The *Internal Revenue Manual* requires that the duties and responsibilities of certain functions be divided and separated among different individuals in order to prevent harmful activities that can result from collusion. This includes dividing mission functions and distinct information system support functions among different individuals or roles.

However, IRS had not corrected a previously identified deficiency by implementing a process to reduce the risk of users being assigned incompatible security roles for one financial system. Specifically, users were still assigned to security roles and other roles that the agency had defined as incompatible for users who have a security role. Until IRS corrects this deficiency, increased risk exists that inadvertent or deliberate misuse of inappropriate privileges may occur on this system.

IRS Developed, Documented, Tested, and Updated Contingency Plans with Results from Testing for All But One Plan

Losing the capability to process, retrieve, and protect electronically maintained information can significantly affect an agency's ability to accomplish its mission. If contingency plans are inadequate, even relatively minor interruptions can result in lost or incorrectly processed data, which can cause financial losses, expensive recovery efforts, and inaccurate or incomplete information. Contingency planning includes developing, testing, and maintaining plans to ensure that when unexpected events occur, critical operations can continue without interruption or can be promptly resumed, and that information resources are protected.

IRS had documented policies for developing and testing information system contingency plans. Specifically, the *Internal Revenue Manual* requires the agency to develop contingency plans for all information systems, and to test the plans to determine their effectiveness and the agency's readiness to execute the plans. The manual also requires the agency to have the capability to continue performance of mission essential functions during any disruption for up to 30 days or until normal operations can resume. Further, the manual requires that IRS review its contingency plans annually and update them to reflect any changes to the agency's information systems and operating environment, and problems encountered during execution or testing.

IRS corrected a previously identified weakness by documenting the extent to which it had capabilities to continue its essential operations. Specifically, the agency documented the disaster recovery steps for switching two different production system platforms to a disaster recovery environment for its payment posting system. In addition, IRS had tested the 10 contingency plans we reviewed.

The agency also had updated 9 of the 10 contingency plans we reviewed to reflect changes to computer equipment and software supporting the information systems and the operating environment. However, it did not fully update one plan to document the existence of a server that was added to the operating environment for one of its tax processing systems. By not updating the contingency plan to reflect the change to the operating environment for one of its tax processing system, IRS has reduced assurance of its ability to fully restore the system in the event of a service interruption. According to IRS, subsequent to our review, the latest version update of the contingency plan for its tax processing system was completed in March 2018; however, we have not yet validated this action, but plan to do so during our fiscal year 2018 audit.

IRS Did Not Consistently Implement Certain Components of Its Information Security Program

An underlying reason for the aforementioned information system security control deficiencies in IRS's financial and tax processing systems was that the agency had not consistently implemented components of a comprehensive information security program. An information security management program should establish a continuous cycle of activity for assessing risk, developing and implementing effective security procedures, and monitoring the effectiveness of these procedures.

In accordance with their responsibilities under FISMA, each agency is required to develop, document, and implement an information security program that, among other things, includes (1) periodic assessments of risk; (2) risk-based policies and procedures; (3) plans for providing adequate information security for networks, facilities, and systems; (4) periodic testing and evaluation of the effectiveness of information security policies, procedures, and practices; and (5) a process for planning, implementing, evaluating, and documenting remedial actions to address information security deficiencies.

IRS Assessed Risk and Identified Some Threats and Vulnerabilities for Selected Systems, but Did Not Correct a Prior Deficiency Affecting Certain Systems

Identifying and assessing information security risks are essential to determining what controls are required to cost-effectively protect information and information systems. Moreover, by increasing awareness of risks, these assessments can generate support for the policies and controls that are adopted as a result. Risk is determined by identifying potential threats to the organization and vulnerabilities in its systems; determining the likelihood that a particular threat may exploit vulnerabilities; and assessing the resulting impact on the organization's mission, including the effect on sensitive and critical systems and data.

IRS had developed and documented policies for identifying, assessing, and managing information security risk to its systems. In this regard, the *Internal Revenue Manual* requires all information systems and data supporting critical operations and assets to be periodically assessed for the risk and magnitude of harm that could result from vulnerabilities and potential threats. The manual also requires that the agency identify and document threats, vulnerabilities, and potential impacts, and review the results periodically.

IRS had conducted and documented risk assessments for the financial systems we reviewed. The risk assessments identified threats, vulnerabilities, and potential impact to the agency's operations. In addition, IRS had updated the assessments within the agency defined frequency of at least 3 years. However, the agency had not corrected a previously identified deficiency where it lacked sufficient justification for its acceptance of the risks associated with making certain configuration decisions in the production environment.[8] Until IRS corrects this weakness, the agency has reduced assurance that its process for managing risk is effectively implemented.

IRS Had Not Yet Corrected Weaknesses in Procedures That Support Components of the Agency-Wide Information Security Program

A key component of an effective information security program is to develop, document, and implement risk-based policies, procedures, and technical standards. These policies, procedures, and technical standards are to govern the security of an agency's computing environment to reduce the risk associated with unauthorized system access or disruption of services.

IRS had developed and documented policies and procedures that addressed several components of its agency-wide information security program. For example, it had documented policies and procedures governing risk assessments, security planning, testing and evaluating information system security controls, and remediating control deficiencies.

Nevertheless, deficiencies that we previously reported in IRS's information security procedures, standards, and guidelines had not been fully corrected. For example, the agency

- did not have detailed procedures for performing reviews of audit records for a financial system,[9] and

---

[8] GAO, *Information Security: Control Deficiencies Continue to Limit IRS's Effectiveness in Protecting Sensitive Financial and Taxpayer Data*, GAO-17-395 (Washington, D.C.: July 26, 2017).

[9] GAO-17-395

- had not updated its configuration standards and guidelines for network devices to incorporate recommendations from industry leaders, security agencies, and key practices from IRS partners to address known vulnerabilities applicable to IRS's environment.[10]

Until it corrects these weaknesses, IRS has limited assurance that its staff will consistently perform reviews of the financial system audit records and configure network devices to effectively protect the agency's information systems.

IRS Developed and Documented Security Plans, but Did Not Always Update Them to Reflect System or Operating Environment Changes

A system security plan provides an overview of the system's security requirements and describes the extent to which security controls are in place or are planned to meet those requirements. The Office of Management and Budget (OMB) requires that agencies develop system security plans for information systems.[11] Further, IRS's *Internal Revenue Manual* requires that the agency's security plans be reviewed at least annually, or as a result of a significant change; and be updated to address changes to the information system and the system's environment of operation. Security plans should also be updated to address problems identified during the plan's implementation or security control assessments.

Although IRS had developed and documented security plans for the 12 systems we reviewed, the agency had not updated 3 of the plans to reflect changes to the information systems or their current operating environment. Specifically, IRS did not update 1 plan to show that the agency had changed the system authentication mechanism to PIV cards, which replaced the weaker encryption that was previously used. In addition, plans for the other 2 systems were not updated to reflect changes in system boundaries, where their interconnections to each other were removed. IRS

---

[10] GAO-17-395.

[11] Office of Management and Budget, *Managing Information as a Strategic Resource,* Circular No. A-130 (Washington, D.C.: July 28, 2016).

also did not correct a similar weakness in the plan for a system that covered multiple sub-systems providing network infrastructure services to the agency, which we reported in fiscal year 2016.[12]

Further, IRS did not update 5 system security plans to remove references to criteria the agency had rescinded. Specifically, IRS rescinded its IT Security Audit Logging Security Standard, effective February 28, 2017. However, at the end of our audit, the plans still referenced the rescinded standard as audit logging criteria.

Without updated system security plans, IRS has less assurance that it has documented and implemented appropriate security controls to protect its sensitive financial and taxpayer information.

Weaknesses Remained in Testing and Evaluating Controls

Another key component of an information security program is the testing and evaluation of controls to determine whether they are effective and operating as intended. IRS's *Internal Revenue Manual* requires management to test and evaluate the effectiveness of information security policies and procedures. It also requires the agency to annually test and evaluate the effectiveness of the security controls in IRS information systems to determine the extent to which the controls are implemented correctly, operating as intended, and producing the desired outcome. The manual further requires that the agency monitor and verify the configuration compliance of mainframe systems using IRS-approved compliance verification applications or the approved security posture monitoring system.

IRS had implemented numerous processes for testing and evaluating its controls to determine whether they were effective and operating as intended. Nevertheless, uncorrected shortcomings continued to exist in the agency's testing and evaluation process. For example, the agency had not

---

[12] GAO, *Information Security: IRS Needs to Further Improve Controls over Financial and Taxpayer Data*, GAO-16-398 (Washington, D.C.: Mar. 28, 2016).

- established test and evaluation procedures to ensure that IRS's control testing methodology and results fully met the intent of the control objectives being tested,[13] or
- addressed limitations with the use of a mainframe tool for verifying compliance with security policies in its mainframe computing environment.[14]

In addition, IRS had not yet corrected a previously reported shortcoming for considering and documenting the results of its review of internal controls related to financial reporting.[15] OMB's *Management's Responsibility for Enterprise Risk Management and Internal Control*, Circular No. A-123, and its related implementation guide (A-123 guide)[16] require an agency's management to monitor and assess controls, including controls over automated information systems that affect financial reporting, and provide an annual assurance statement on the overall adequacy and effectiveness of internal control within the agency. The A-123 guide further specifies that a service organization's systems are considered to be part of an entity's information system.[17] In addition, IRS's documented procedures for the review of external systems that support the agency's financial reporting require that IRS review, when available, those systems' service organization controls (SOC) reports.[18]

---

[13] GAO, *Information Security: IRS Needs to Continue Improving Controls over Financial and Taxpayer Data*, GAO-15-337 (Washington, D.C.: Mar. 19, 2015).

[14] GAO-17-395.

[15] GAO-17-395.

[16] Office of Management and Budget, *Management's Responsibility for Enterprise Risk Management and Internal Control*, Circular No. A-123, is the policy document that implements the requirements of 31 U.S.C. 3512 (c), (d) (commonly known as the *Federal Managers' Financial Integrity Act* or FMFIA). Circular No. A-123's focus for internal controls is primarily on providing agencies with a framework for assessing and managing risks more strategically and effectively. The circular was recently revised to reflect changes incorporated in GAO's updated *Standards for Internal Control in the Federal Government*.

[17] Agencies are responsible for assessing the extent to which they rely on the internal controls of its service organization and, where appropriate, monitoring the effectiveness of internal control over its financial reporting at service organizations.

[18] American Institute of Certified Public Accountants, Auditing Standards Board, Attestation Standards: Clarification and Recodification, AT-C Section 320 Reporting on an Examination of Controls at a Service Organization Relevant to User Entities' Internal Control Over Financial Reporting, Statement on Standards for Attestation Engagements No. 18 (April 2016), contains performance and reporting requirements and application guidance

Although IRS had reviewed the SOC reports received for external systems used for financial reporting, it had not identified and documented the user controls that it deemed relevant to its internal control environment. Since the agency had not identified the relevant user controls, it had not tested the operating effectiveness for those controls. We reported the same deficiency regarding the SOC reports for two of the same external systems in fiscal year 2016.[19]

Without identifying, verifying, and reviewing user controls, IRS has limited assurance that it has the appropriate controls in place or will draw adequate conclusions on the operating effectiveness of these controls. Because of the shortcomings in its process for testing and evaluating controls, IRS may not be fully aware of vulnerabilities that could adversely affect its critical applications and data.

Deficiencies in IRS's Remediation Process Remained

As part of an information security program, agencies are required to have a process for planning, implementing, evaluating, and documenting remedial actions to address deficiencies in information security policies, practices, and procedures. The *Internal Revenue Manual* requires that IRS, among other things, track the status and resolution of all weaknesses and verify that each weakness is corrected before closing them.

Although IRS had a process in place for tracking and implementing remedial actions to resolve known control deficiencies, it was not always effective in verifying whether the remedial actions were successfully implemented. For example, at the beginning of the fiscal year 2017 audit, the agency informed us that it had implemented 63 of the 166 recommendations that we made during prior audits. However, we determined that it had effectively implemented only 37 (about 59 percent) of these 63 recommendations.

---

for a service auditor examining controls at organizations that provide services to user entities when those controls are likely to be relevant to user entities' internal control over financial reporting.

[19] GAO-17-395.

We also concluded that 7 (of the original 166) recommendations were no longer relevant due to the changes in IRS's operating environment.[20] Further, we found that an additional 5 recommendations (of the original 166) that IRS had not submitted to us for validation, had been adequately addressed by the agency. Collectively, this indicated that the agency had corrected or mitigated deficiencies associated with 49 of the 166 recommendations to resolve control weaknesses that were open at the beginning of the audit.

Although IRS made some progress in correcting or mitigating the previously reported deficiencies, the agency still had not fully or effectively implemented corrective actions for 117—about 70 percent—of the 166 recommendations. This indicates that IRS's remedial action verification process continues to be ineffective, although we previously made a recommendation to the agency to improve its process for verifying the effectiveness of actions to remedy deficiencies.[21] Until IRS takes additional steps to implement a more effective process, it will have limited assurance that control deficiencies are being properly mitigated or corrected.

CONCLUSION

During fiscal year 2017, IRS continued to make progress in addressing deficiencies in internal control and had successfully addressed a number of our prior recommendations concerning information system security control deficiencies. However, continuing and newly identified control deficiencies limited the effectiveness of security controls for protecting the confidentiality, integrity, and availability of the agency's financial and tax processing systems. As a result, sensitive financial and taxpayer data on IRS computer systems will remain vulnerable until the agency addresses

---

[20] These 7 recommendations no longer reflect IRS's current operating environment and will either be reissued to more closely align with the agency's current policies and environment or not be reissued due to being covered by another recommendation.

[21] GAO-15-337.

the deficiencies for which we previously made 117 recommendations, as well as the newly identified deficiencies we highlight in this chapter in the areas of access control, configuration management, segregation of duties, contingency planning, and security management.

The collective effect of these deficiencies in information security is the basis of our determination that IRS had a significant deficiency in internal control over financial reporting systems as of September 30, 2017. Continued and consistent management commitment and attention to an effective information security program will be essential to the maintenance of, and continued improvements in, the agency's information security controls.

RECOMMENDATIONS FOR EXECUTIVE ACTION

We are making the following 5 recommendations to IRS:

The Commissioner of Internal Revenue should take steps to improve the implementation of IRS's information security program by entering correct contractor password expiration dates, per IRS's policy, in the system used for managing user access authorizations. (Recommendation 1)

The Commissioner of Internal Revenue should take steps to improve the implementation of IRS's information security program by documenting access authorizations for non-unique accounts. (Recommendation 2)

The Commissioner of Internal Revenue should take steps to improve the implementation of IRS's information security program by reviewing non-unique accounts at least annually, per IRS's policy. (Recommendation 3)

The Commissioner of Internal Revenue should take steps to improve the implementation of IRS's information security program by updating security plans for three systems to reflect changes to their operating environment. (Recommendation 4)

The Commissioner of Internal Revenue should take steps to improve the implementation of IRS's information security program by removing from five systems security plans, references to logging standards that IRS has rescinded. (Recommendation 5)

We are also making 32 technical recommendations in a separate report with limited distribution. These recommendations address information system security control deficiencies related to identification and authentication, audit and monitoring, configuration management, and contingency planning.

AGENCY COMMENTS AND OUR EVALUATION

We received written comments on a draft of this chapter from IRS. In its comments, reproduced in appendix II, the agency agreed with our recommendations and stated that it plans to review them carefully and ensure that its corrective actions include root cause analysis for sustainable fixes that implement appropriate security controls. According to the agency, it is committed to improving its financial management, internal controls, information technology security posture, and the overall effectiveness of its information system controls.

IRS also asserted that the integrity of its financial systems continues to be sound and that the agency has enhanced its ability to protect and defend against malicious acts and expanded its use of continuous application security monitoring and fraud prevention and detection. However, as we noted in this chapter, although the agency has continued to make progress in addressing information security control deficiencies, it has not always effectively implemented access, and other controls to protect the confidentiality, integrity, and availability of its financial systems and information. The effective implementation of our recommendations in this chapter and in our previous reports will assist IRS in protecting taxpayer and financial information.

Sincerely yours,

Nancy R. Kingsbury
Managing Director, Applied Research and Methods

Gregory C. Wilshusen
Director, Information Security Issues

APPENDIX I: OBJECTIVE, SCOPE, AND METHODOLOGY

Our objective for this audit was to assess whether IRS's controls over selected financial and tax processing systems were effective in ensuring the confidentiality, integrity, and availability of sensitive financial and taxpayer data.

To determine whether controls over selected financial and tax processing systems were effective, we considered the results of our evaluation of IRS's actions to mitigate previously reported control deficiencies and performed new audit work at two IRS enterprise computing centers located in Martinsburg, West Virginia, and Memphis, Tennessee, as well as IRS facilities in Detroit, Michigan, and New Carrollton, Maryland. We concentrated our evaluation on threats emanating from sources internal to IRS's computer networks. Considering systems that directly or indirectly support the processing of material transactions that are reflected in the agency's financial statements, we focused our work on systems and applications that directly or indirectly support financial and taxpayer information systems.

Our evaluation was based on GAO's *Federal Information System Controls Audit Manual*,[22] which contains guidance for reviewing information system controls that affect the confidentiality, integrity, and availability of computerized information; National Institute of Standards and Technology guidance; and IRS policies, procedures, practices, and standards. We evaluated controls by

---

[22] GAO, *Federal Information System Controls Audit Manual* (FISCAM), GAO-09-232G (Washington, D.C.: February 2009).

- reviewing configurations on IRS's network devices to determine if implemented configurations would protect the devices against malicious code and unauthorized access;
- comparing the complexity, expiration, and other settings for passwords on systems and databases to IRS and federal guidelines to determine whether password strength requirements were being enforced;
- evaluating whether access to systems and databases were appropriately limited, according to IRS policy and federal and vendor best practices;
- examining IRS's implementation of cryptography to secure data transmissions in order to determine whether implemented cryptographic mechanisms met the requirements of applicable federal standards;
- analyzing audit logs of events occurring in system environments responsible for taxpayer data processing and the support of refunds disbursements, revenue, unpaid assessments, and payroll financial reporting;
- observing and reviewing physical security controls at both enterprise computing centers to determine whether computer facilities and resources were protected from espionage, sabotage, damage, and theft;
- evaluating the mainframe configuration controls supporting financial management processing;
- evaluating the access controls over disk storage shared across multiple mainframe processing environments;
- evaluating the access controls of the mainframe operating systems that support payroll and taxpayer data processing;
- comparing security configurations on systems and databases to IRS and federal guidelines;
- comparing the release dates of vendor-supplied software components to the install dates of the software running on IRS's systems to ensure that software was up to date; and

- reviewing continuity of operations plans to determine whether they contained the details necessary for the recovery of system and business functions, and assessing the extent to which those details had been documented and tested.

Using the requirements in the *Federal Information Security Modernization Act of 2014*,[23] which established components of an agency-wide information security program, we evaluated IRS's implementation of its security program by

- reviewing risk assessments to determine whether they were being updated within the agency defined frequency of at least three years;
- reviewing IRS's policies, procedures, practices, and standards to determine whether its security management program had been documented, approved, and was up to date;
- reviewing IRS's system security plans for selected systems to determine the extent to which the plans had been reviewed and included information as required by the National Institute of Standards and Technology;
- examining documentation to determine the extent to which IRS was performing internal control reviews of financial systems;
- analyzing documentation to determine whether the effectiveness of security controls had been periodically assessed;
- reviewing IRS's actions to correct previously reported control deficiencies to determine whether the agency had effectively mitigated or resolved the control deficiencies; and

---

[23] The Federal Information Security Modernization Act of 2014 (FISMA 2014) (Pub. L. No. 113-283, Dec. 18, 2014) largely superseded the Federal Information Security Management Act of 2002 (FISMA 2002), enacted as Title III, E-Government Act of 2002, Pub. L. No. 107-347, 116 Stat. 2899, 2946 (Dec. 17, 2002). As used in this report, FISMA refers to the new requirements in FISMA 2014, and to other relevant FISMA 2002 requirements that were unchanged by FISMA 2014 and continue in full force and effect.

- reviewing continuity of operations plans for selected systems to determine whether such plans had been appropriately documented and tested.

In addition, we discussed with management officials and key security representatives, including those from IRS's Computer Security Incident Response Center and Information Technology Cybersecurity organization, as well as the two computing centers, whether information system security controls were in place, adequately designed, and operating effectively.

We performed our work in accordance with generally accepted government auditing standards. Those standards require that we plan and perform the audit to obtain sufficient appropriate evidence to provide a reasonable basis for our findings and conclusions based on our audit objectives. We believe that the evidence obtained provides a reasonable basis for our findings and conclusions based on our audit objective.

APPENDIX II: COMMENTS FROM THE INTERNAL REVENUE SERVICE

DEPARTMENT OF THE TREASURY
INTERNAL REVENUE SERVICE
WASHINGTON, D.C. 20224

COMMISSIONER

July 12, 2018

Mr. Gregory C. Wilshusen
Director, Information Security Issues
U.S. Government Accountability Office
441 G Street, N.W.
Washington, DC 20548

Dear Mr. Wilshusen:

Thank you for the opportunity to comment on the draft report titled, IRS Needs to Rectify Control Deficiencies That Limit Its Effectiveness in Protecting Sensitive Financial and Taxpayer Data, *GAO-18-391* (public version).

We are pleased the Government Accountability Office (GAO) recognized our progress in addressing a number of Information Technology (IT) security controls. We also appreciate your acknowledging that the IT organization is relied upon extensively to effectively achieve the IRS mission. IRS processed 9.1 million returns on opening day of Filing Season 2018, with an average of 160 federal acknowledgements every second. IRS received more than 119 million tax returns, answered more than 23 million taxpayer questions on our toll-free help lines and provided in-person assistance to more than 860,000 people. IRS.gov has been visited more than 363 million times this year. These accomplishments are even more remarkable given that Congress enacted 30 retroactive tax law changes in early February 2018 and the IRS implemented these changes in the middle of the filing season.

In FY 2017, our increased effort to address GAO's prior recommendations resulted in successful implementation and closure of 42 prior year recommendations compared to 19 in FY 2016. We anticipate this improvement will continue in FY 2018 due to our proactive implementation of 81 additional prior year recommendations. We feel this proactive approach has also contributed to the decreased number of recommendations from 98 in FY 2016 to 38 in FY 2017.

We appreciate the recommendations and are aware that much work remains. We agree with and are reviewing each of your recommendations carefully and will ensure our corrective actions include root cause analysis for sustainable fixes that implement appropriate security controls within our technology and human capital resource limitations, and we will provide the detailed corrective action plans in our 60-day letter response to Congress. We are also reviewing all prior year open recommendations to ensure they continue to be relevant in our current environment.

As you know, the IRS is committed to improving its financial management, internal controls, information technology security posture and the overall effectiveness of its information system controls. We have enhanced our ability to protect and defend against malicious acts and expanded our use of continuous application security monitoring and fraud prevention and detection. The continued security and privacy of all taxpayer information is of the utmost importance to us, and the integrity of our financial systems continues to be sound. We appreciate your continued support and guidance as we work to address the recommendations and look forward to working with you to develop appropriate measures.

Sincerely,

David J. Kautter
Acting Commissioner

In: Information Security
Editor: Paul Scholz

ISBN: 978-1-53615-880-9

Chapter 3

INFORMATION SECURITY: SUPPLY CHAIN RISKS AFFECTING FEDERAL AGENCIES*

Gregory C. Wilshusen

WHY GAO DID THIS STUDY

IT systems are essential to the operations of the federal government. The supply chain—the set of organizations, people, activities, and resources that create and move a product from suppliers to end users— for IT systems is complex and global in scope. The exploitation of vulnerabilities in the IT supply chain is a continuing threat. Federal security guidelines provide for managing the risks to the supply chain.

This testimony statement highlights information security risks associated with the supply chains used by federal agencies to procure IT

* This is an edited, reformatted and augmented version of United States Government Accountability Office, Testimony before the Subcommittees on Counterterrorism and Intelligence, and Oversight and Management Efficiency, Committee on Homeland Security, House of Representatives, Publication No. GAO-18-667T, dated July 12, 2018.

systems. The statement also summarizes GAO's 2012 report that assessed the extent to which four national security-related agencies had addressed such risks. To develop this statement, GAO relied on its previous reports, as well as information provided by the national security-related agencies on their actions in response to GAO's previous recommendations. GAO also reviewed federal information security guidelines and directives.

What GAO Recommends

In its 2012 report, GAO recommended that Justice, Energy, and DHS take eight actions, as needed, to develop and document policies, procedures, and monitoring capabilities that address IT supply chain risk. The departments generally concurred with the recommendations and subsequently implemented seven recommendations and partially implemented the eighth recommendation.

What GAO Found

Reliance on a global supply chain introduces multiple risks to federal information systems. Supply chain threats are present during the various phases of an information system's development life cycle and could create an unacceptable risk to federal agencies. Information technology (IT) supply chain-related threats are varied and can include:

- installation of intentionally harmful hardware or software (i.e., containing "malicious logic");
- installation of counterfeit hardware or software;
- failure or disruption in the production or distribution of critical products;
- reliance on malicious or unqualified service providers for the performance of technical services; and

- installation of hardware or software containing unintentional vulnerabilities, such as defective code.

These threats can have a range of impacts, including allowing adversaries to take control of systems or decreasing the availability of materials needed to develop systems. These threats can be introduced by exploiting vulnerabilities that could exist at multiple points in the supply chain. Examples of such vulnerabilities include the acquisition of products or parts from unauthorized distributors; inadequate testing of software updates and patches; and incomplete information on IT suppliers. Malicious actors could exploit these vulnerabilities, leading to the loss of the confidentiality, integrity, or availability of federal systems and the information they contain.

GAO reported in 2012 that the four national security-related agencies in its review—the Departments of Defense, Justice, Energy, Homeland Security (DHS)—varied in the extent to which they had addressed supply chain risks. Of the four agencies, Defense had made the most progress addressing the risks. It had defined and implemented supply chain protection controls, and initiated efforts to monitor the effectiveness of the controls. Conversely, Energy and DHS had not developed or documented policies and procedures that defined security measures for protecting against IT supply chain threats and had not developed capabilities for monitoring the implementation and effectiveness of the measures. Although Justice had defined supply chain protection measures, it also had not developed or documented procedures for implementing or monitoring the measures.

Energy and Justice fully implemented the recommendations that GAO made in its 2012 report and resolved the deficiencies that GAO had identified with their supply chain risk management efforts by 2016. DHS also fully implemented two recommendations to document policies and procedures for defining and implementing security measures to protect against supply chain threats by 2015, but could not demonstrate that it had fully implemented the recommendation to develop and implement a monitoring capability to assess the effectiveness of the security measures.

Chairmen King and Perry, Ranking Members Rice and Correa, and Members of the Subcommittees:

Thank you for the opportunity to testify at today's hearing on keeping adversaries away from the homeland security supply chain. As you know, federal agencies and the owners and operators of our nation's critical infrastructure rely extensively on information technology (IT) and IT services to carry out their operations. Securing this technology, its supply chain, and the information it contains is essential to protecting national and economic security.

Since 1997, we have identified federal information security as a governmentwide high-risk area. In 2003, we expanded this high-risk area to include protecting systems supporting our nation's critical infrastructure.[1]

My statement provides an overview of the information security risks associated with the supply chains used by federal agencies to procure IT equipment, software, or services.[2] The statement also discusses our 2012 assessment of the extent to which four national security-related agencies—the Departments of Defense, Justice, Energy, and Homeland Security (DHS)—had addressed these risks.[3]

In developing this testimony, we relied on our previous reports,[4] as well as information provided by the national security-related agencies on their actions in response to our previous recommendations. We also considered information contained in special publications issued by the

---

[1] See, most recently, GAO, *High-Risk Series: Progress on Many High-Risk Areas, While Substantial Efforts Needed on Others,* GAO-17-317 (Washington, D.C.: Feb. 15, 2017).

[2] The National Institute of Standards and Technology (NIST) has defined the term "supply chain" as a set of organizations, people, activities, information, and resources that create and move a product or service from suppliers to an organization's customers. NIST defines "information technology" as any equipment or interconnected system or subsystem of equipment that is used in the automatic acquisition, storage, manipulation, management, movement, control, display, switching, interchange, transmission, or reception of data or information. This includes, among other things, computers, software, firmware, and services (including support services).

[3] GAO, *IT Supply Chain: National Security-Related Agencies Need to Better Address Risks,* GAO-12-361 (Washington, D.C.: Mar. 23, 2012).

[4] See GAO-12-361; *State Department Telecommunications: Information on Vendors and Cyber-Threat Nations,* GAO-17-688R (Washington, D.C.: July 27, 2017); and *Telecommunications Networks: Addressing Potential Security Risks of Foreign-Manufactured Equipment*, GAO-13-625T (Washington, D.C.: May 21, 2013).

National Institute of Standards and Technology (NIST) and a directive issued by DHS. A more detailed discussion of the objectives, scope, and methodology for this work is included in each of the reports that are cited throughout this statement.

The work on which this statement is based was conducted in accordance with generally accepted government auditing standards. Those standards require that we plan and perform audits to obtain sufficient, appropriate evidence to provide a reasonable basis for our findings and conclusions. We believe that the evidence obtained provided a reasonable basis for our findings and conclusions based on our audit objectives.

BACKGROUND

The design and development of information systems can be complex undertakings, consisting of a multitude of pieces of equipment and software products, and service providers. Each of the components of an information system may rely on one or more supply chains—that is, the set of organizations, people, activities, information, and resources that create and move a product or service from suppliers to an organization's customers.

Obtaining a full understanding of the sources of a given information system can also be extremely complex. According to the Software Engineering Institute, the identity of each product or service provider may not be visible to others in the supply chain. Typically, an acquirer, such as a federal agency, may only know about the participants to which it is directly connected in the supply chain. Further, the complexity of corporate structures, in which a parent company (or its subsidiaries) may own or control companies that conduct business under different names in multiple countries, presents additional challenges to fully understanding the sources of an information system. As a result, the acquirer may have little visibility into the supply chains of its suppliers.

Federal procurement law and policies promote the acquisition of commercial products when they meet the government's needs. Commercial providers of IT use a global supply chain to design, develop, manufacture, and distribute hardware and software products throughout the world. Consequently, the federal government relies heavily on IT equipment manufactured in foreign nations.

Table 1. Possible Manufacturing Locations of Typical Network Components

Component	Possible manufacturing locations
Workstations	United States, Israel, Spain, China, Malaysia, Singapore, United Kingdom
Notebook computers	United States, Israel, Spain, China, Malaysia, Singapore, United Kingdom
Routing and switching	United States, India, Belgium, Canada, China, Germany, Israel, Japan, Netherlands, Poland, United Kingdom
Fiber optic cabling	China, Malaysia, Vietnam, Japan, Thailand, Indonesia
Servers	Brazil, Canada, United States, India, Japan, France, Germany, United Kingdom, Israel, Singapore
Printers	Japan, United States, Germany, France, Netherlands, Taiwan, China, Malaysia, Thailand, Vietnam, Philippines

Source: GAO analysis of public information | GAO-18-667T.

Federal information and communications systems can include a multitude of IT equipment, products, and services, each of which may rely on one or more supply chains. These supply chains can be long, complex, and globally distributed and can consist of multiple tiers of outsourcing. As a result, agencies may have little visibility into, understanding of, or control over how the technology that they acquire is developed, integrated, and deployed, as well as the processes, procedures, and practices used to ensure the integrity, security, resilience, and quality of the products and services. Table 1 highlights possible manufacturing locations of typical components of a computer or information systems network.

Moreover, many of the manufacturing inputs required for these components—whether physical materials or knowledge—are acquired from various sources around the globe. Figure 1 depicts the potential countries of origin of common suppliers of various components in a commercially available laptop computer.

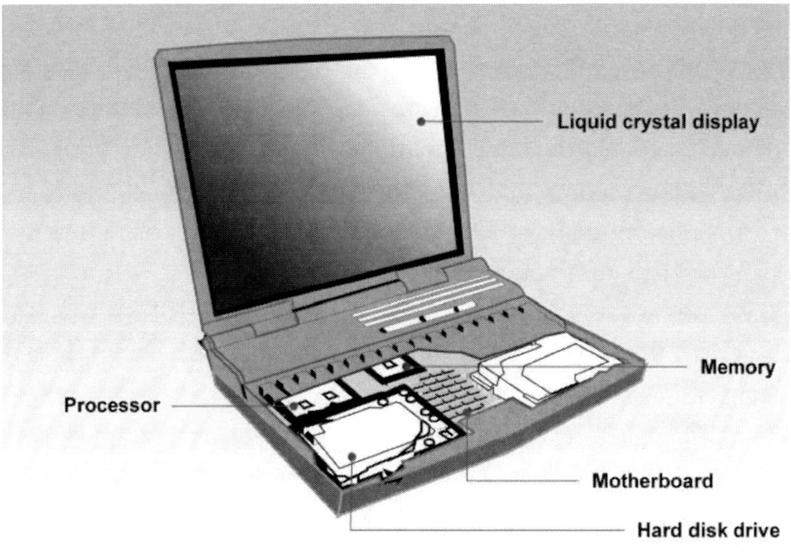

Component	Location of facilities potentially used by suppliers
Liquid crystal display	China, Czech Republic, Japan, Poland, Singapore, Slovak Republic, South Korea, Taiwan
Memory	China, Israel, Italy, Japan, Malaysia, Philippines, Puerto Rico, Singapore, South Korea, Taiwan, United States
Processor	Canada, China, Costa Rica, Ireland, Israel, Malaysia, Singapore, United States, Vietnam
Motherboard	Taiwan
Hard disk drive	China, Ireland, Japan, Malaysia, Philippines, Singapore, Thailand, United States

Source: GAO analysis of public information | GAO-18-667T.

Figure 1. Potential Origins of Common Suppliers of Laptop Components.

Federal Laws and Guidelines Require the Establishment of Information Security Programs and Provide for Managing Supply Chain Risk

The *Federal Information Security Modernization Act* (FISMA) of 2014 requires federal agencies to develop, document, and implement an agency-wide information security program to provide information security for the information systems and information that support the operations and assets of the agency.[5] The act also requires that agencies ensure that information security is addressed throughout the life cycle of each agency information system. FISMA assigns NIST the responsibility for providing standards and guidelines on information security to agencies. In addition, the act authorizes DHS to develop and issue binding operational directives to agencies, including directives that specify requirements for the mitigation of exigent risks to information systems.

NIST has issued several special publications (SP) that provide guidelines to federal agencies on controls and activities relevant to managing supply chain risk. For example,

- NIST SP 800-39 provides an approach to organization-wide management of information security risk, which states that organizations should monitor risk on an ongoing basis as part of a comprehensive risk management program.[6]
- NIST SP 800-53 (Revision 4) provides a catalogue of controls from which agencies are to select controls for their information systems. It also specifies several control activities that organizations could use to provide additional supply chain protections, such as conducting due diligence reviews of suppliers and developing acquisition policy, and implementing procedures that help protect against supply chain threats throughout the system development life cycle.[7]
- NIST SP 800-161 provides guidance to federal agencies on identifying, assessing, selecting, and implementing risk management processes and mitigating controls throughout their

[5] FISMA 2014 (Pub. L. No. 113-283, Dec. 18, 2014) largely superseded the *Federal Information Security Management Act of 2002* (FISMA 2002), enacted as *Title III, E-Government Act of 2002*, Pub. L. No. 107-347, 116 Stat. 2899, 2946 (Dec. 17, 2002). As used in this statement, FISMA refers both to FISMA 2014 and to those provisions of FISMA 2002 that were either incorporated into FISMA 2014 or were unchanged and continue in full force and effect.

[6] NIST, *Managing Information Security Risk: Organization, Mission, and Information System View*, SP 800-39 (Gaithersburg, Md.: March 2011).

[7] NIST, *Security and Privacy Controls for Federal Information Systems and Organizations*, SP 800-53, Revision 4 (Gaithersburg, Md.: April 2013).

organizations to help manage information and communications technology supply chain risks.[8]

In addition, as of June 2018, DHS has issued one binding operational directive related to an IT supply chain-related threat. Specifically, in September 2017, DHS issued a directive to all federal executive branch departments and agencies to remove and discontinue present and future use of Kaspersky-branded products on all federal information systems.[9] In consultation with interagency partners, DHS determined that the risks presented by these products justified their removal.

Beyond these guidelines and requirements, the *Ike Skelton National Defense Authorization Act for Fiscal Year 2011* also included provisions related to supply chain security. Specifically, Section 806 authorizes the Secretaries of Defense, the Army, the Navy, and the Air Force to exclude a contractor from specific types of procurements on the basis of a determination of significant supply chain risk to a covered system.[10] Section 806 also establishes requirements for limiting disclosure of the basis of such procurement action.

IT Supply Chains Introduce Numerous Information Security Risks to Federal Agencies

In several reports issued since 2012,[11] we have pointed out that the reliance on complex, global IT supply chains introduces multiple risks to federal information and telecommunications systems. This includes the risk of these systems being manipulated or damaged by leading foreign

---

[8] NIST, *Supply Chain Risk Management Practices for Federal Information Systems and Organizations*, SP-800-161 (Gaithersburg, Md.: April 2015).

[9] DHS, *Removal of Kaspersky-Branded Products*, BOD-17-01 (Washington, D.C.: Sept. 13, 2017).

[10] The act defines "supply chain risk" as "risk that an adversary may sabotage, maliciously introduce unwanted function, or otherwise subvert the design, integrity, manufacturing, production, distribution, installation, operation, or maintenance of a covered system so as to surveil, deny, disrupt, or otherwise degrade the function, use, or operation of such system."

[11] GAO-12-361, GAO-13-652T, and GAO-17-688R.

cyber-threat nations such as Russia, China, Iran, and North Korea.[12] Threats and vulnerabilities created by these cyber-threat nations, vendors or suppliers closely linked to cyber-threat nations,[13] and other malicious actors can be sophisticated and difficult to detect and, thus, pose a significant risk to organizations and federal agencies.

As we reported in March 2012,[14] supply chain threats are present at various phases of a system's development life cycle. Key threats that could create an unacceptable risk to federal agencies include the following.

- Installation of hardware or software containing malicious logic, which is hardware, firmware, or software that is intentionally included or inserted in a system for a harmful purpose. Malicious logic can cause significant damage by allowing attackers to take control of entire systems and, thereby, read, modify, or delete sensitive information; disrupt operations; launch attacks against other organizations' systems; or destroy systems.
- Installation of counterfeit hardware or software, which is hardware or software containing non-genuine component parts or code. According to the Defense Department's Information Assurance

[12] The Office of the Director of National Intelligence has identified Russia, China, Iran, and North Korea as leading cyber-threat nations in its *Worldwide Threat Assessment of the U.S. Intelligence Community* (Washington, D.C.: Feb. 9, 2016 and Feb. 13, 2018).

[13] The Department of State Authorities Act, Fiscal Year 2017, defines "closely-linked" as, with respect to a foreign supplier, contactor, or subcontractor and a cyber-threat nation, (1) incorporated or headquartered in the territory; (2) having ties to the military forces; (3) having ties to the intelligence services; or (4) the beneficiary of significant low-interest or no-interest loans, loan forgiveness, or other support of a leading cyber-threat nation. The Act also included a provision for GAO to review the Department of State's (State) critical telecommunications equipment or services obtained from manufacturers or suppliers that are closely linked to the leading cyber-threat nations. Based on GAO's open source review of generalizable samples of 52 telecommunications device manufacturers and software developers supporting the State's critical telecommunications capabilities and 100 of State's telecommunications contractors, GAO identified 16 companies--12 equipment manufacturers and software developers and 4 telecommunications contractors--with suppliers reported to be headquartered in cyber-threat nations. All of these suppliers were reported to be headquartered in China or, in one case, Russia. The data did not establish whether State's telecommunications capabilities were supported by equipment or software originating from suppliers linked to companies in GAO's samples. GAO did not identify any reported military ties, intelligence ties, or low-interest loans involving cyber-threat nations among any of the suppliers. See GAO-17-688R.

[14] GAO-12-361.

Technology Analysis Center, counterfeit IT threatens the integrity, trustworthiness, and reliability of information systems for several reasons, including the facts that (1) counterfeits are usually less reliable and, therefore, may fail more often and more quickly than genuine parts; and (2) counterfeiting presents an opportunity for the counterfeiter to insert malicious logic or backdoors[15] into replicas or copies that would be far more difficult in more secure manufacturing facilities.[16]

- Failure or disruption in the production or distribution of critical products. Both man-made (e.g., disruptions caused by labor, trade, or political disputes) and natural (e.g., earthquakes, fires, floods, or hurricanes) causes could decrease the availability of material needed to develop systems or disrupt the supply of IT products critical to the operations of federal agencies.
- Reliance on a malicious or unqualified service provider for the performance of technical services. By virtue of their position, contractors and other service providers may have access to federal data and systems. Service providers could attempt to use their access to obtain sensitive information, commit fraud, disrupt operations, or launch attacks against other computer systems and networks.
- Installation of hardware or software that contains unintentional vulnerabilities, such as defects in code that can be exploited. Cyber attackers may focus their efforts on, among other things, finding and exploiting existing defects in software code. Such defects are usually the result of unintentional coding errors or misconfigurations, and can facilitate attempts by attackers to gain unauthorized access to an agency's information systems and data, or disrupt service.

---

[15] A "backdoor" is a general term for a malicious program that can potentially give an intruder remote access to an infected computer.

[16] Information Assurance Technology Analysis Center, *Security Risk Management for the Off-the-Shelf (OTS) Information and Communications Technology (ICT) Supply Chain, An Information Assurance Technology Analysis Center State of the Art Repor*t, DO 380 (Herndon, Va.: August 2010).

We noted in the March 2012 report that threat actors[17] can introduce these threats into federal information systems by exploiting vulnerabilities that could exist at multiple points in the global supply chain. In addition, supply chain vulnerabilities can include weaknesses in agency acquisition or security procedures, controls, or implementation related to an information system. Examples of the types of vulnerabilities that could be exploited include

- acquisitions of IT products or parts from sources other than the original manufacturer or authorized reseller, such as independent distributors, brokers, or on the gray market;
- lack of adequate testing for software updates and patches; and
- incomplete information on IT suppliers.

If a threat actor exploits an existing vulnerability, it could lead to the loss of the confidentiality, integrity, or availability of the system and associated information. This, in turn, can adversely affect an agency's ability to carry out its mission.

Four National Security-Related Agencies Have Acted to Better Address IT Supply Chain Risks for Their Information Systems

In March 2012, we reported that the four national security-related agencies (i.e., Defense, Justice, Energy, and DHS) had acknowledged the risks presented by supply chain vulnerabilities.[18] However, the agencies varied in the extent to which they had addressed these risks by (1) defining supply chain protection measures for department information systems, (2) developing implementing procedures for these measures, and (3)

[17] Supply chain-related threat actors include foreign intelligence services and militaries, corporate spies, corrupt government officials, cyber vandals, disgruntled employees, radical activists, purveyors of counterfeit goods, or criminals.

[18] GAO-12-361.

establishing capabilities for monitoring compliance with, and the effectiveness of, such measures.

Of the four agencies, the Department of Defense had made the most progress addressing the risks. Specifically, the department's supply chain risk management efforts began in 2003 and included:

- a policy requiring supply chain risk to be addressed early and across a system's entire life cycle and calling for an incremental implementation of supply chain risk management through a series of pilot projects;
- a requirement that every acquisition program submit and update a "program protection plan" that was to, among other things, help manage risks from supply chain exploits or design vulnerabilities;
- procedures for implementing supply chain protection measures, such as an implementation guide describing 32 specific measures for enhancing supply chain protection and procedures for program protection plans identifying ways in which programs should manage supply chain risk; and
- a monitoring mechanism to determine the status and effectiveness of supply chain protection pilot projects, as well as monitoring compliance with and effectiveness of program protection policies and procedures for several acquisition programs.

Conversely, our report noted that the other three agencies had made limited progress in addressing supply chain risks for their information systems. For example:

- The Department of Justice had defined specific security measures for protecting against supply chain threats through the use of provisions in vendor contracts and agreements. Officials identified (1) a citizenship and residency requirement and (2) a national security risk questionnaire as two provisions that addressed supply chain risk. However, Justice had not developed procedures for ensuring the effective implementation of these protection measures or a mechanism for verifying compliance with, and the effectiveness of these measures. We stressed that, without such procedures, Justice would have limited assurance that its departmental information systems were being adequately protected against supply chain threats.

- In May 2011, the Department of Energy revised its information security program, which required Energy components to implement provisions based on NIST and Committee on National Security Systems guidance. However, the department was unable to provide details on implementation progress, milestones for completion, or how supply chain protection measures would be defined. Because it had not defined these measures or associated implementing procedures, we reported that the department was not in a position to monitor compliance or effectiveness.
- Although its information security guidance mentioned the NIST control related to supply chain protection, DHS had not defined the supply chain protection control activities that system owners should employ. The department's information security policy manager stated that DHS was in the process of developing policy that would address supply chain protection, but did not provide details on when it would be completed. In the absence of such a policy, DHS was not in a position to develop implementation procedures or to monitor compliance or effectiveness.

To assist Justice, Energy, and DHS in better addressing IT supply chain-related security risks for their departmental information systems, we made eight recommendations to these three agencies in our 2012 report. Specifically, we recommended that Energy and DHS:

- develop and document departmental policy that defines which security measures should be employed to protect against supply chain threats.

We also recommended that Justice, Energy, and DHS:

- develop, document, and disseminate procedures to implement the supply chain protection security measures defined in departmental policy, and
- develop and implement a monitoring capability to verify compliance with, and assess the effectiveness of, supply chain protection measures.

The three agencies generally agreed with our recommendations and, subsequently, implemented seven of the eight recommendations. Specifically, we verified that Justice and Energy had implemented each of the recommendations we made to them by 2016. We also confirmed that DHS had implemented two of the three recommendations we made to that agency by 2015.

However, as of fiscal year 2016,[19] DHS had not fully implemented our recommendation to develop and implement a monitoring capability to verify compliance with, and assess the effectiveness of, supply chain protections. Although the department had developed a policy and approach for monitoring supply chain risk management activities, it could not provide evidence that its components had actually implemented the policy. Thus, we were not able to close the recommendation as implemented. Nevertheless, the implementation of the seven recommendations and partial implementation of the eighth recommendation better positioned the three agencies to monitor and mitigate their IT supply chain risks.

In addition, we reported in March 2012 that the four national security-related agencies had participated in interagency efforts to address supply chain security, including participation in the Comprehensive National Cybersecurity Initiative,[20] development of technical and policy tools, and

---

[19] GAO reviews agency actions to implement its recommendations and may decide to close a recommendation as not implemented if an agency has not implemented the recommendation within 4 fiscal years of GAO making the recommendation. Fiscal year 2016 was the fourth fiscal year after GAO made the recommendations to DHS in its March 2012 report.

[20] Begun by the Bush administration in 2008, the Comprehensive National Cybersecurity Initiative is a series of initiatives aimed at improving cybersecurity within the federal

collaboration with the intelligence community. In support of the cybersecurity initiative, Defense and DHS jointly led an interagency initiative on supply chain risk management to address issues of globalization affecting the federal government's IT. Also, DHS had developed a comprehensive portfolio of technical and policy-based product offerings for federal civilian departments and agencies, including technical assessment capabilities, acquisition support, and incident response capabilities. The efforts of the four agencies could benefit all federal agencies in addressing their IT supply chain risks.

In summary, the global IT supply chain introduces a myriad of security risks to federal information systems that, if realized, could jeopardize the confidentiality, integrity, and availability of federal information systems. Thus, the potential exists for serious adverse impact on an agency's operations, assets, and employees. These factors highlight the importance and urgency of federal agencies appropriately assessing, managing, and monitoring IT supply chain risk as part of their agencywide information security programs.

Chairmen King and Perry, Ranking Members Rice and Correa, and Members of the Subcommittees, this completes my prepared statement. I would be pleased to answer your questions.

---

government. This initiative, which is composed of 12 projects with the objective of safeguarding federal executive branch information systems, includes a project focused on addressing global supply chain risk management.

In: Information Security
Editor: Paul Scholz
ISBN: 978-1-53615-880-9

Chapter 4

INFORMATION SECURITY: OPM HAS IMPLEMENTED MANY OF GAO'S 80 RECOMMENDATIONS, BUT OVER ONE-THIRD REMAIN OPEN*

United States Government Accountability Office

November 13, 2018

The Honorable James Lankford
Chairman

The Honorable Christopher Coons
Ranking Member
Subcommittee on Financial Services and General Government
Committee on Appropriations
United States Senate

---

* This is an edited, reformatted and augmented version of the United States Government Accountability Office, A Briefing for Staff of the Senate and House Appropriations Subcommittees on Financial Services and General Government, Publication No. GAO-19-143R, dated September 26, 2018.

The Honorable Tom Graves
Chairman

The Honorable Mike Quigley
Ranking Member
Subcommittee on Financial Services and General Government
Committee on Appropriations
United States House of Representatives

The Office of Personnel Management (OPM) collects and maintains personal data on millions of individuals, including data related to security clearance investigations. In June 2015, OPM reported that an intrusion into its systems had affected the personnel records of about 4.2 million current and former federal employees. Then, in July 2015, the agency reported that a separate but related incident had compromised its systems and the files related to background investigations for 21.5 million individuals.

From February 2015 through August 2017, we conducted multiple reviews of OPM's information security. We issued four reports based on these reviews.[1] The reports contained 80 recommendations for improving the agency's security posture.

The Explanatory Statement that accompanies the *Consolidated Appropriations Act, 2018*, included a provision for GAO to brief the House and Senate Appropriations Committees on actions taken by OPM in response to GAO's information security recommendations.[2] Our specific objective was to determine the extent to which OPM has implemented our recommendations to improve the agency's information security.

To accomplish this objective, we reviewed relevant documents and artifacts reflecting OPM's actions and progress toward implementing the

---

[1] GAO, *Information Security: Agencies Need to Improve Controls over Selected High-Impact Systems*, GAO-16-501 (Washington, D.C.: May 18, 2016); *Information Security: OPM Needs to Improve Controls over Selected High-Impact Systems*, GAO-16-687SU (Washington, D.C.: Aug. 15, 2016); *Information Security: OPM Has Improved Controls, but Further Efforts Are Needed*, GAO-17-459SU (Washington, D.C.: Aug. 3, 2017); and *Information Security: OPM Has Improved Controls, but Further Efforts Are Needed*, GAO-17-614 (Washington, D.C.: Aug. 3, 2017).

[2] Explanatory Statement accompanying Pub. L. No. 115-141, Div. E, Title V (2018), 164 Cong. Rec. H2045, H2522 (March 22, 2018).

80 recommendations contained in the four reports, and assessed the actions against the intent of the recommendations. Our review examined the agency's actions taken on our recommendations through September 20, 2018. We also evaluated the agency's remedial action plans for those recommendations not yet implemented to determine the agency's intentions regarding the outstanding recommendations. Further, we interviewed officials in OPM's Office of the Chief Information Officer to obtain additional clarification about the agency's actions, as needed.

On September 26, 2018, we briefed members of your staff on the results of our review. This chapter formally transmits the final briefing slides (see enc. I).

We conducted this performance audit from September 2018 through November 2018 in accordance with generally accepted government auditing standards. Those standards require that we plan and perform the audit to obtain sufficient, appropriate evidence to provide a reasonable basis for our findings and conclusions based on our audit objective. We believe that the evidence obtained provides a reasonable basis for our findings and conclusions based on our audit objective.

Table 1. OPM's Implementation of GAO's Information Security Program and Control Recommendations, as of September 20, 2018

GAO Report Number	Number of Recommendations			
	Closed–implemented	Open–insufficient evidence	Open– no evidence	Total
GAO-16-501	0	1	3	4
GAO-16-687SU	46	2	14	62
GAO-17-459SU	2	1	6	9
GAO-17-614	3	1	1	5
Total	51	5	24	80

Source: GAO analysis of OPM evidence. I GAO-19-143R.

Notes: Closed-implemented: GAO validated that OPM implemented the recommendation.

Open-insufficient evidence: GAO determined that evidence provided by OPM was insufficient to demonstrate that the agency had implemented the recommendation.

Open-no evidence: OPM did not provide GAO with any evidence that the agency had implemented the recommendation.

In summary, OPM has made progress in implementing our recommendations for improving its security posture, but further actions are needed. As of September 20, 2018, the agency had implemented 51 (about 64 percent) of the 80 recommendations, but had not provided any evidence, or provided insufficient evidence, to demonstrate implementation of the remaining recommendations, as shown in Table 1.

The following provides further detail on the status of the recommendations, as of September 20, 2018, for each of the four reports:

- GAO-16-501 - OPM had not provided sufficient evidence to demonstrate that it had implemented the 4 recommendations. Three of the recommendations were related to enhancing security plans, performing comprehensive security control assessments, and updating remedial action plans for two selected high-impact systems.[3] The fourth recommendation was to provide and track specialized training for all individuals, including contractors, who have significant security responsibilities. We designated 3 of the 4 recommendations as priority recommendations.[4]
- GAO-16-687SU – Of the 62 recommendations GAO made in this chapter, OPM had implemented or addressed 46 recommendations,[5] including those recommendations associated with strengthening firewall controls, enforcing password policies, restricting access to a key server, logging security-related activities, and updating the contingency plan for a high-impact system. However, the agency had not provided sufficient evidence

[3] Systems that agencies categorize as high impact are those systems where the loss of confidentiality, integrity, or availability could have a severe or catastrophic adverse effect on organizational operations, assets, or individuals.

[4] Priority recommendations are those that GAO believes warrant priority attention from heads of departments and agencies. In a July 2016 letter to OPM's Acting Director, the Comptroller General informed her of these three priority recommendations related to strengthening controls for high-impact systems. In June 2017, the Comptroller General sent a similar letter highlighting these three priority recommendations to another OPM Acting Director, as well as in another letter to the OPM Director in March 2018 because the agency had not implemented the recommendations.

[5] OPM addressed 21 of the 46 recommendations through the decommissioning of one of the two high-impact systems that we reviewed.

that it had implemented the other 16 recommendations. These recommendations included avoiding the use of the same administrator accounts by multiple persons, implementing procedures governing the use of special privileges on a key computer, encrypting passwords while stored or in-transit across the network, and installing the latest versions of operating system software on network devices supporting a high-impact system.

- GAO-17-459SU – OPM had implemented 2 of the 9 recommendations we made in this chapter. In response to the 2 recommendations, the agency improved its protection of data by encrypting data for two selected systems we reviewed. However, the agency had not demonstrated that it had fully implemented 7 other recommendations, including recommendations to reset all passwords subsequent to the breach, install critical patches in a timely manner, periodically evaluate accounts to ensure privileged access is warranted, and assess controls on selected systems as defined in its continuous monitoring plan.
- GAO-17-614 - OPM had implemented 3 of the 5 recommendations in this chapter. The agency had updated milestones for completing United States Computer Emergency Readiness Team recommendations[6] and its policy for deploying threat indicators, and had improved its guidance for evaluating the quality of control assessments. However, the agency had not demonstrated that it had implemented 2 recommendations that we had designated as priority recommendations[7] to improve the timeliness of validating corrective actions and to develop and implement training requirements for staff using special tools.

---

[6] The United States Computer Emergency Readiness Team (US-CERT), a component of the Department of Homeland Security, operates the federal information security incident center. In September 2015, US-CERT made 19 recommendations to OPM to bolster the agency's information security practices and controls in the wake of the 2015 breaches.

[7] Based on the need to strengthen controls over information technology systems, in a March 2018 letter to the OPM Director, the Comptroller General cited these two additional priority recommendations from our August 2017 report.

According to officials in OPM's Office of the Chief Information Officer, the agency plans to implement 25 of the remaining 29 open recommendations by the end of calendar year 2018. The agency expects to implement 3 additional recommendations by the end of fiscal year 2019. OPM has created remedial action plans for each of the 28 open recommendations that it plans to implement.

However, the officials stated that the agency does not plan to implement the one remaining recommendation related to deploying a security tool on contractor workstations. The agency asserted that it has compensating controls in place to address the intent of this recommendation, but has not provided evidence to us of these controls. Expeditiously implementing all remaining open recommendations is essential to ensuring that appropriate controls are in place to protect the agency's systems and information.

Agency Comments

We provided a draft of this chapter to the Acting Director of the Office of Personnel Management. The agency had no additional comments beyond what is reprinted in the enclosed briefing slides.

We are sending copies of this chapter to the appropriate congressional committees, the Acting Director of OPM, the OPM Office of the Inspector General, and other interested congressional parties.

Gregory C. Wilshusen
Director, Information Security Issues

Background

- The Office of Personnel Management (OPM) collects and maintains personal data on millions of individuals, including data related to security clearance investigations.

- In June 2015, OPM reported that an intrusion into its information systems had compromised the personnel records of about 4.2 million current and former federal employees. Then, in July 2015, the agency reported that a separate, but related, incident had compromised its systems and the data files related to background investigations for 21.5 million individuals.

The *Federal Information Security Modernization Act of 2014* requires agencies to implement an information security program that includes, among other things,

- periodic assessments of the risk and magnitude of harm that could result from the unauthorized access, use, disclosure, disruption, modification, or destruction of information or information systems;
- plans for providing adequate information security for networks and
- systems;
- training personnel with significant security responsibilities for information
- security;
- periodic testing and evaluation of the effectiveness of information security
- policies, procedures, and practices; and
- a process for planning, implementing, evaluating, and documenting remedial action to address any deficiencies in the information security policies, procedures, or practices of the agency.

Government-wide information security requirements include, among other things:

- tightening policies and practices for privileged users by inventorying and minimizing the number of privileged users,

- when the assessed risk indicates the need, encrypting federal information at rest and in transit unless otherwise protected by alternative physical and logical safeguards, and
- developing and maintaining a strategy for monitoring information security controls on an ongoing and continuous basis.
- From February 2015 through August 2017, we conducted multiple reviews of OPM's information security. We issued four reports based on these reviews. The reports contained 80 recommendations for improving the agency's information security posture.
- The Explanatory Statement that accompanies the *Consolidated Appropriations Act, 2018,* included a provision for GAO to brief the House and Senate Appropriations Subcommittees on Financial Services and General Government on actions taken by OPM in response to GAO's information security recommendations.[8]

Objective

Our objective was to:

- Determine the extent to which OPM has implemented GAO's recommendations to improve the agency's information security.

Scope and Methodology

We examined OPM's implementation of the 80 recommendations contained in the following four reports:

---

[8] Explanatory Statement to accompany Pub. L. No. 115-141, Div. E, Title V (2018), 164 Cong. Rec. H2045, H2522 (March 22, 2018).

1. *Information Security: Agencies Need to Improve Controls over Selected High-Impact Systems*, GAO-16-501, May 18, 2016.
2. *Information Security: OPM Needs to Improve Controls over Selected High-Impact Systems*, GAO-16-687SU, August 15, 2016.
3. *Information Security: OPM Has Improved Controls, but Further Efforts Are Needed*, GAO-17-459SU, August 3, 2017.
4. *Information Security: OPM Has Improved Controls, but Further Efforts Are Needed*, GAO-17-614, August 3, 2017 (public version of GAO-17-459SU).

We considered the corrective actions that OPM had taken as of September 20, 2018.

To accomplish the objective, we:

- reviewed relevant documents and artifacts reflecting OPM's actions and progress toward implementing our recommendations, and assessed the agency's actions against the intent of the recommendations;
- evaluated the agency's remedial action plans for recommendations not yet implemented to determine the agency's intentions regarding these recommendations; and
- interviewed officials in OPM's Office of the Chief Information Officer to obtain clarification on the actions taken to implement the recommendations, as needed.

We conducted this performance audit in September 2018 in accordance with generally accepted government auditing standards.

Those standards require that we plan and perform the audit to obtain sufficient, appropriate evidence to provide a reasonable basis for our findings and conclusions based on our audit objective. We believe that the evidence obtained provides a reasonable basis for our findings and conclusions based on our audit objective.

Results in Brief

- OPM has made progress in implementing GAO's information security recommendations, but further efforts remain. As of September 20, 2018, the agency had implemented 51 of the 80 recommendations.
- For the 29 open recommendations, OPM has not provided any evidence, or insufficient evidence, to demonstrate implementation.
- OPM intends to implement 25 of the 29 open recommendations by the end of 2018 and 3 of them by the end of fiscal year 2019. The agency does not intend to implement one of the recommendations to deploy a security tool on contractor workstations. OPM stated it had compensating controls, but has not provided evidence of them.
- Expeditiously implementing the open recommendations is essential to ensuring appropriate controls are in place to protect the agency's systems and information.

Prior Findings and Current Status of Recommendations

Table 2. OPM's implementation of GAO's information security program and control recommendations, as of September 20, 2018

	Number of Recommendations			
GAO Report	Closed – implemented	Open – insufficient evidence provided	Open – no evidence provided	Total
GAO-16-501	0	1	3	4
GAO-16-687SU	46	2	14	62
GAO-17-459SU	2	1	6	9
GAO-17-614	3	1	1	5
Totals	51	5	24	80

Source: GAO analysis of OPM evidence.

As of September 20, 2018, OPM had implemented many recommendations; nonetheless, over one-third remain open because the agency has not yet provided any evidence, or has provided insufficient evidence, to demonstrate that it has taken the actions we recommended. See Table 2.

Prior Findings and Current Status of Recommendations – GAO-16-501

In a May 2016 report on federal agencies' implementation of controls over their high-impact systems,[9] we noted that weaknesses existed in the two OPM high-impact information systems that we reviewed.

These weaknesses existed, in part, because OPM had not effectively implemented elements of its information security program.

Although the agency had developed risk assessments for the two selected systems and a continuous monitoring strategy that included performance metrics, it had not taken other steps.

For example:

- Although OPM had identified almost all controls required for high-impact systems in the security plan for one of the selected systems, the agency had identified only about half of the high-impact controls in the other system's plan.
- OPM also had not performed comprehensive security control assessments. For example, in performing its assessments, OPM had not identified the access control weaknesses that we identified during our review.
- The agency had not included updated milestones in its remedial action plans.
- The agency was not tracking specialized training for all individuals with significant security responsibilities.
- We made 4 recommendations, including 3 priority recommendations,[10] to enhance security plans, perform

---

[9] Systems that agencies categorize as high impact are those systems where the loss of confidentiality, integrity, or availability can have a severe or catastrophic adverse effect on organizational operations, assets, or individuals.

comprehensive security control assessments, and update remedial action plans for selected systems, and track specialized training.

- OPM concurred with 2 of the 4 recommendations (security and remedial action plans), partially concurred with a third (specialized training), and did not concur with a fourth recommendation (security control assessments). OPM subsequently developed remedial action plans to implement all 4 recommendations.
- As of September 20, 2018, OPM had not provided sufficient evidence to demonstrate that it had implemented any of the 4 recommendations.

Prior Findings and Current Status of Recommendations – GAO-16-687SU

In an August 2016 report, we noted that the agency had established controls for the two high-impact systems we reviewed. However, we pointed out that OPM had not always effectively implemented these controls in a manner to protect the confidentiality, integrity, and availability of the systems. Specifically, the agency had not adequately controlled access to the systems.

For example, OPM had not always

- protected the selected systems' boundaries,[11]
- enforced password policies for authenticating access to the systems,
- restricted access to only that needed to perform job duties,
- enabled encryption for a database, and

[10] Priority recommendations are those that GAO believes warrant priority attention from heads of departments and agencies.

[11] Boundary protection controls pertain to the protection of a logical boundary around a system by implementing measures to prevent unauthorized information exchange across the boundary in either direction.

- provided sufficient logging for auditing and monitoring the systems.

Weaknesses also existed in configuration management and contingency planning for the two selected systems.

- We made 62 recommendations to OPM to mitigate weaknesses that we identified in access controls, configuration management, and contingency planning for the two systems.
- OPM concurred with the recommendations for one of the two systems. At the time that we issued our report, the agency stated that the other system, which was contractor-operated, had been disconnected, and that the agency would evaluate the recommendations for that system when a new contract was in place.
- Subsequently, in December 2016, OPM stated that it had contracted with a vendor, and that the vendor was actively working to address our recommendations for the second system.

As of September 20, 2018, OPM had implemented or addressed[12] 46 of the 62 recommendations, including those recommendations associated with:

- strengthening firewall controls;
- enforcing password policies;
- restricting access to a key server;
- logging security-related activities; and
- updating the contingency plan for a high-impact system.

---

[12] OPM addressed 21 of the 46 recommendations through the decommissioning of one of the two high-impact systems that we reviewed.

However, OPM had not provided sufficient evidence that it had implemented the other 16 recommendations. These recommendations included:

- avoiding the use of the same administrator accounts by multiple people;
- implementing procedures governing the use of special privileges on a key computer;
- encrypting passwords while stored or in-transit across the network;

and

- installing the latest versions of operating system software on network devices supporting a high-impact system.

OPM had developed remedial action plans for each of the 16 open recommendations.

Prior Findings and Current Status of Recommendations – GAO-17-459SU

In an August 2017 report on OPM's efforts to improve its security program, we noted that the agency had implemented or made progress toward implementing 19 recommendations that the United States Computer Emergency Readiness Team (US-CERT)[13] had previously made to bolster OPM's information security practices and controls in the wake of the 2015 breaches.

Specifically, OPM had implemented 11 of the US-CERT recommendations and had partially implemented the remaining 8

[13] The United States Computer Emergency Readiness Team, a component of DHS, operates the federal information security incident center.

recommendations, with actions for 4 of the 8 requiring further improvement.

- We reported that OPM had also made progress in implementing information security policies and practices associated with certain government-wide requirements. However, it had not fully implemented all of the requirements.
- We made 9 recommendations for OPM to improve upon the actions it had taken to implement the US-CERT recommendations and to further implement government-wide requirements.

OPM concurred with 7 of our recommendations, partially concurred with 1 recommendation, and had developed remedial action plans to implement all 8 of these recommendations.

- The agency did not concur with, and does not plan to implement, our recommendation related to deploying a security tool on contractor workstations. OPM asserted that it had compensating controls in place, but has not yet provided sufficient evidence of these controls.

As of September 20, 2018, OPM had improved its protection of data by implementing 2 of our recommendations related to encrypting data for two of the selected systems that we reviewed.

However, the agency had not demonstrated that it had fully implemented the other 6 recommendations for which it concurred or partially concurred, including recommendations to

- reset all passwords subsequent to the breach;
- install critical patches in a timely manner;
- periodically evaluate accounts to ensure privileged access is warranted; and
- assess controls on selected systems as defined in its continuous monitoring plan.

Prior Findings and Current Status of Recommendations – GAO-17-614

In our fourth report, issued in August 2017, we noted that, although OPM had made progress toward implementing the US-CERT recommendations and addressing government-wide requirements, the agency had not

- validated remedial actions related to the recommendations in a timely manner or consistently updated milestones for outstanding recommendations;
- complied with its own plan for conducting periodic control assessments;
- documented role-based training requirements for individuals configuring and maintaining monitoring tools; and
- always comprehensively tested key security controls on selected contractor-operated systems.

We made 5 recommendations, including 2 priority recommendations, for OPM to further improve its security posture. These recommendations were related to

- updating milestones for completing the US-CERT recommendations,
- validating remedial actions more timely,
- updating policy for deploying threat indicators,
- developing and implementing training requirements for staff using monitoring tools, and
- providing guidance on evaluating the quality of control assessments.

- OPM concurred with 4 of the 5 recommendations, partially concurred with 1 recommendation, and had developed remedial action plans to implement all 5 recommendations.
- As of September 20, 2018, OPM had implemented 3 of the recommendations. Specifically, the agency had updated (1) milestones for completing US-CERT recommendations and (2) its policy for deploying threat indicators. The agency had also improved its guidance for evaluating the quality of control assessments.
- However, the agency could not demonstrate that it had implemented the 2 priority recommendations to improve the timeliness of validating corrective actions, and to develop and implement training requirements for staff using monitoring tools.

Prior Findings and Current Status of Recommendations

- According to officials in OPM's Office of the Chief Information Officer, OPM plans to implement 25 of the remaining 29 open recommendations by the end of 2018. The agency expects to implement 3 recommendations by the end of fiscal year 2019.
- As previously noted, OPM does not plan to implement the 1 remaining recommendation to install a security tool on contractor workstations.

GAO SUMMARY

- Implementing all of the remaining open recommendations expeditiously is essential to OPM ensuring that appropriate security controls are in place and operating as intended.

- Until OPM implements these recommendations, its systems and information will be at increased risk of unauthorized access, use, disclosure, modification, or disruption.

AGENCY COMMENTS

In providing written comments on a draft of these slides, OPM stated the following:

- During FY 2018, OPM made significant progress addressing GAO recommendations.
- In FY 2018, the agency committed significant resources to address GAO audit recommendations.
- This commitment has resulted in noteworthy progress, demonstrated by the implementation of 51 of the recommendations in FY 2018 alone.
- OPM is dedicated to continued implementation of the remaining recommendations, projecting implementation of 5 more recommendations during this final quarter of FY18.
- OPM will build on the progress of FY 2018 with planned implementation of 20 recommendations in Q1 of FY 2019 and the remaining 3 recommendations in Q4 of FY 2019.

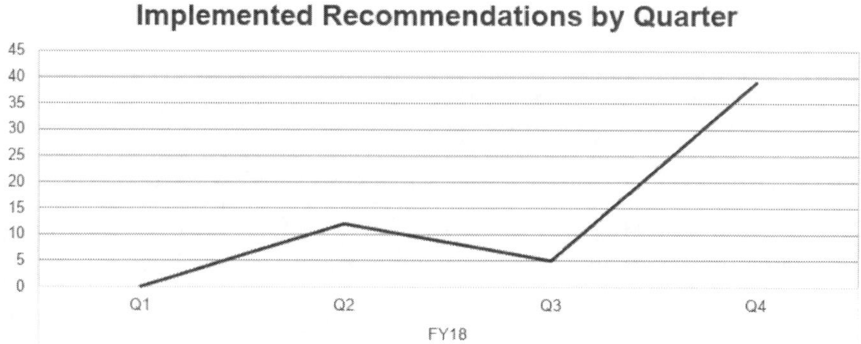

OPM also provided the graph above regarding the number of GAO information security recommendations implemented by quarter in fiscal year 2018.

In: Information Security
Editor: Paul Scholz

ISBN: 978-1-53615-880-9

Chapter 5

INFORMATION SECURITY: SIGNIFICANT PROGRESS MADE, BUT CDC NEEDS TO TAKE FURTHER ACTION TO RESOLVE CONTROL DEFICIENCIES AND IMPROVE ITS PROGRAM*

United States Government Accountability Office

ABBREVIATIONS

CDM	continuous diagnostics and mitigation
CIO	chief information officer
CISO	chief information security officer
CDC	Centers for Disease Control and Prevention
DMZ	demilitarized zone

* This is an edited, reformatted and augmented version of United States Government Accountability Office; Report to Congressional Requesters; Accessible Version, Publication No. GAO-19-70, dated December 2018.

ISCM	information security continuous monitoring
FIPS	Federal Information Processing Standards
FISMA	Federal Information Security Modernization Act of 2014
HHS	Department of Health and Human Services
ISC	Interagency Security Committee
ISSO	information systems security officer
NIST	National Institute of Standards and Technology
OCIO	Office of the Chief Information Officer
PIV	personal identity verification
POA&M	plan of actions and milestones
SP	special publication

Why GAO Did This Study

CDC is responsible for detecting and responding to emerging health threats and controlling dangerous substances. In carrying out its mission, CDC relies on information technology systems to receive, process, and maintain sensitive data. Accordingly, effective information security controls are essential to ensure that the agency's systems and information are protected from misuse and modification.

GAO was asked to examine information security at CDC. In June 2018, GAO issued a limited official use only report on the extent to which CDC had effectively implemented technical controls and an information security program to protect the confidentiality, integrity, and availability of its information on selected information systems.

This current report is a public version of the June 2018 report. In addition, for this public report, GAO determined the extent to which CDC has taken corrective actions to address the previously identified security program and technical control deficiencies and related recommendations for improvement. For this chapter, GAO reviewed supporting documents regarding CDC's actions on previously identified recommendations and interviewed personnel at CDC.

What GAO Found

As GAO reported in June 2018, the Centers for Disease Control and Prevention (CDC) implemented technical controls and an information security program that were intended to safeguard the confidentiality, integrity, and availability of its information systems and information. However, GAO identified control and program deficiencies in the core security functions related to identifying risk, protecting systems from threats and vulnerabilities, detecting and responding to cyber security events, and recovering system operations (see table below). GAO made 195 recommendations to address these deficiencies.

As of August 2018, CDC had made significant progress in resolving many of the security deficiencies by implementing 102 of 184 (about 55 percent) technical control recommendations, and partially implementing 1 of 11 information security program recommendations made in the June 2018 report. The figure shows the status of CDC's efforts to implement the 195 recommendations.

Number of GAO-Identified Technical Control and Information Security Program Deficiencies at the Centers for Disease Control and Prevention and Associated Recommendations by Core Security Function

Core security function	Number of technical control deficiencies	Number of technical control recommendations	Number of information security program deficiencies	Number of information security program recommendations
Identify	0	0	5	5
Protect	85	161	1	1
Detect	8	18	3	3
Respond	1	5	1	1
Recover	0	0	1	1
Total	94	184	11	11

Source: GAO. | GAO-19-70.

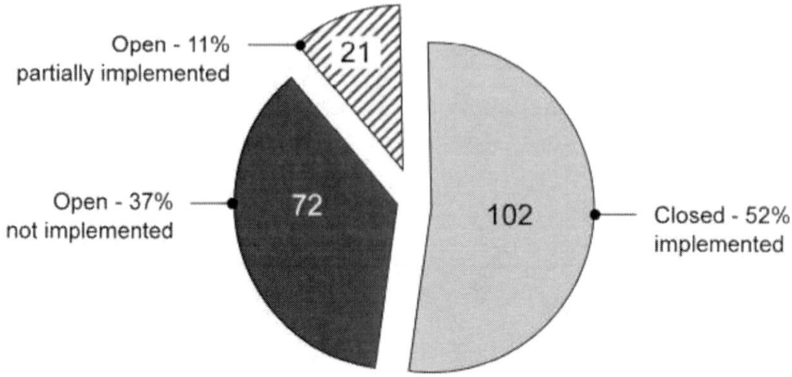

Source: GAO analysis of CDC data. | GAO-19-70.

Status of GAO Recommendations to the Centers for Disease Control and Prevention.

Additionally, CDC has created remedial action plans to implement the majority of the remaining open recommendations by September 2019. Until CDC implements these recommendations and resolves the associated deficiencies, its information systems and information will remain at increased risk of misuse, improper disclosure or modification, and destruction.

December 20, 2018

The Honorable Greg Walden
Chairman
Committee on Energy and Commerce
House of Representatives

The Honorable Trey Gowdy
Chairman

The Honorable Elijah E. Cummings
Ranking Member
Committee on Oversight and Government Reform
House of Representatives

The Honorable Fred Upton
House of Representatives

The Centers for Disease Control and Prevention (CDC) is responsible for protecting America from both foreign and domestic health, safety, and security threats. Its roles include detecting and responding to new and emerging health threats, tackling the biggest health problems causing death and disability for Americans, putting science and advanced technology into action to prevent disease, and controlling dangerous and exotic substances that can cause incurable and deadly diseases. Performing these roles may involve tracking diseases and finding out what is making people sick and the most effective ways to prevent diseases.

CDC confronts global disease threats through advanced computing and laboratory analysis of large amounts of data to find solutions, making use of computer systems critical to the process. However, cyber incidents at federal agencies demonstrate the damage that increasingly sophisticated threats can cause and underscore the importance of effectively protecting federal systems, including those used by CDC to achieve its mission.

Since 1997, we have designated the security of information on federal systems (i.e., information security) to be a government-wide high-risk area. In 2003, we expanded the area to include securing the computerized systems supporting the nation's critical infrastructure and, in 2015, we included protecting the privacy of personally identifiable information.[1]

Given the critical role that CDC performs and concerns over the security of federal information systems, you requested that we examine the security controls over key CDC systems. Accordingly, our specific objective was to assess the extent to which CDC had effectively implemented an information security program and controls to protect the confidentiality, integrity, and availability of its information on selected information systems.

---

[1] For our latest high-risk report, see GAO, *High-Risk Series: Progress on Many High-Risk Areas, While Substantial Efforts Needed on Others*, GAO-17-317 (Washington, D.C.: Feb. 15, 2017).

In June 2018, we issued a report that addressed the extent to which CDC had effectively implemented an information security program and controls to protect the confidentiality, integrity, and availability of its information on selected information systems.[2] In the report, we made 184 recommendations to CDC to resolve the technical security control deficiencies in the information systems we reviewed and 11 additional recommendations to improve its information security program. We designated that report as "limited official use only" (LOUO) and did not release it to the general public because of the sensitive information it contained.

This subsequent report publishes the findings discussed in our June 2018 report, but we have removed all references to the sensitive information. Specifically, we deleted the names of the information systems and computer networks that we examined, disassociated identified control deficiencies from named systems, deleted certain details about information security controls and control deficiencies, and omitted an appendix that was contained in the LOUO report. The appendix contained sensitive details about the technical security control deficiencies in CDC's information systems and computer networks that we reviewed, and the 184 recommendations we made to mitigate those deficiencies. We also provided a draft of this chapter to CDC officials to review and comment on the sensitivity of the information contained herein and to affirm that the report can be made available to the public without jeopardizing the security of CDC's information systems and networks.

In addition, this chapter addresses a second objective that was not included in the June 2018 report. Specifically, this objective was to determine the extent to which CDC has taken corrective actions to address the previously identified security program and technical control deficiencies and related recommendations for improvement that we identified in the earlier report.

As noted in our LOUO report, to accomplish the first objective—to determine the extent to which CDC had effectively implemented an

---

[2] GAO, *Information Security: CDC Needs to Improve Its Program and Resolve Control Deficiencies*, GAO-18-437SU (Washington, D.C.: June 20, 2018).

information security program and controls—we had examined the agency's security policies, procedures, and practices, and evaluated the technical security controls over 24 CDC systems. These included 10 key systems, 8 of which were mission-essential, that (1) collect, process, and maintain private or potentially sensitive proprietary business, medical, and personally identifiable information; (2) are essential to CDC's mission; (3) could have a catastrophic or severe impact on operations if compromised; or (4) could be of particular interest to potential adversaries. We also selected 14 general support systems that were part of the network infrastructure supporting the 10 systems.

To review CDC's information security program, we had examined security policies, procedures, and other documents; analyzed risk assessments, security plans, security control assessments, remedial action plans, and contingency plans for 8 selected mission-essential systems; and interviewed personnel at CDC headquarters. To review controls over the 10 key systems and 14 general support systems, we had examined the agency's network infrastructure and assessed the controls associated with system access, encryption, configuration management, and logging and monitoring. We conducted site visits to two CDC facilities located in Atlanta, Georgia.

To accomplish our second objective—on CDC's actions to address the previously identified security program and technical control deficiencies and related recommendations—we requested that the agency provide a status report of its actions to implement each recommendation we made in the June 2018 report. For each recommendation that CDC indicated it had implemented as of August 3, 2018, we examined supporting documents, observed or tested the associated security control or procedure, and/or interviewed the responsible agency officials to assess the effectiveness of the actions taken to implement the recommendation or otherwise resolve the underlying control deficiency. Based on this assessment and CDC status reports, we categorized the status of each recommendation as being closed-implemented, open-partially implemented, or open-not implemented. Additional details on our objectives, scope, and methodology are provided in appendix I.

We conducted this performance audit from December 2016 to December 2018 in accordance with generally accepted government auditing standards. Those standards require that we plan and perform the audit to obtain sufficient, appropriate evidence to provide a reasonable basis for our findings and conclusions based on our audit objectives. We believe that the evidence obtained provides a reasonable basis for our findings and conclusions based on our audit objectives.

Background

CDC—an operating division of the Department of Health and Human Services (HHS)—serves as the national focal point for disease prevention and control, environmental health, and promotion and education activities designed to improve the health of Americans. The agency is also responsible for leading national efforts to detect, respond to, and prevent illnesses and injuries that result from natural causes or the release of biological, chemical, or radiological agents.

To achieve its mission and goals, the agency relies on an array of partners, including public health associations and state and local public health agencies. It collaborates with these partners on initiatives such as monitoring the public's health, investigating disease outbreaks, and implementing prevention strategies. The agency also uses its staff located in foreign countries to aid in international efforts, such as guarding against global diseases. Table 1 describes the organization of CDC.

CDC is staffed by approximately 20,000 employees across the United States and around the world. For fiscal year 2017, according to agency officials, the agency's total appropriation was approximately $12 billion, of which it reported spending approximately $424 million on information technology. In addition, the officials stated that approximately $31 million (or about 7.3 percent of the amount spent on information technology) was for information security across all CDC information technology investments.

Table 1. Organization of the Centers for Disease Control and Prevention (CDC)

CDC Organization	Description
Office of the Director	Guides agency priorities and activities; coordinates program, science, policy, and communications across CDC; and coordinates administrative management activities.
Deputy Director for Infectious Diseases	Leads, promotes, and facilitates science, programs, and policies to reduce the burden of infectious diseases, both domestically and globally. This office includes the National Center for Emerging and Zoonotic Infectious Diseases; the National Center for HIV/AIDS, Viral Hepatitis, Sexually Transmitted Diseases, and Tuberculosis Prevention; and the National Center for Immunization and Respiratory Diseases.
Deputy Director for Non-Infectious Diseases	Reduces the burden of non-infectious diseases, injuries, birth defects, disabilities, and environmental health hazards. This office includes the National Center on Birth Defects and Developmental Disabilities, the National Center for Chronic Disease Prevention and Health Promotion, the National Center for Environmental Health/Agency for Toxic Substances and Disease Registry, and the National Center for Injury Prevention and Control.
Deputy Director for Public Health Science and Surveillance	Leads, promotes, and facilitates science, surveillance, standards, and policies to reduce the burden of diseases, both domestically and globally. This office includes the Office of Science; the Office of Laboratory Science and Safety; the Center for Surveillance, Epidemiology, and Laboratory Services; and the National Center for Health Statistics.
Deputy Director for Public Health Service and Implementation Science	Leads, promotes, and facilitates science, programs and policies to identify and respond to public health threats, both domestically and internationally. It includes the Office of Minority Health and Health Equity; the Center for Global Health; the Center for Preparedness and Response; and the Center for State, Tribal, Local, and Territorial Support.
National Institute for Occupational Safety and Health	Provides leadership to prevent workplace injuries and illness by conducting scientific research, developing guidance and recommendations, sharing information, and responding to requests for workplace health hazard evaluations.

Source: CDC. | GAO-19-70.

CDC Relies on Information Systems to Help Achieve Its Mission

CDC relies extensively on information technology to fulfill its mission and support related administrative needs. Among the approximately 750 systems reported in its inventory, the agency has systems dedicated to supporting public health science, practice, and administration. All of these systems rely on an information technology infrastructure that includes network components, critical servers, and data centers.

CDC Has Defined Organizational Security Roles and Responsibilities

At CDC, the chief information officer (CIO) is responsible for establishing and enforcing policies and procedures protecting information resources. The CIO is to lead the efforts to protect the confidentiality, integrity, and availability of the information and systems that support the agency and its operations, and is to report quarterly to the HHS CIO on the overall effectiveness of CDC's information security and privacy program, including the progress of remedial actions.

The CIO designated a chief information security officer (CISO), who is to oversee compliance with applicable information security and privacy requirements of the agency. The CISO, among other things, is responsible for providing information security protections commensurate with the risk and magnitude of the harm resulting from unauthorized access, use, disclosure, and disruption of information and information systems that support the operations and assets of the agency.

To further ensure information security compliance, information systems security officers (ISSO) are responsible for managing the information security program within their respective organizations and report on security program matters to the CISO, including computer security-related incidents. ISSO responsibilities include ensuring that vendor-issued security patches are expeditiously installed and that system owners establish processes for timely removal of access privileges when a user's system access is no longer necessary. In addition, security stewards are to perform operational security analyses supporting the efforts of the ISSO. Further, business stewards serve as program managers, accepting full accountability for the operations of the systems and ensuring that security is planned, documented, and properly resourced for each aspect of the information security program.

Federal Laws and Guidance Establish Security Requirements to Protect Federal Information and Systems

The *Federal Information Security Modernization Act* (FISMA) of 2014[3] provides a comprehensive framework for ensuring the effectiveness of information security controls over information resources that support federal operations and assets. FISMA assigns responsibility to the head of each agency for providing information security protections commensurate with the risk and magnitude of the harm resulting from unauthorized access, use, disclosure, disruption, modification, or destruction of information systems used or operated by an agency or by a contractor of an agency or other organization on behalf of an agency. The law also delegates to the agency CIO (or comparable official) the authority to ensure compliance with FISMA requirements. The CIO is responsible for designating a senior agency information security officer whose primary duty is information security.

The law also requires each agency to develop, document, and implement an agency-wide information security program to provide risk-based protections for the information and information systems that support the operations and assets of the agency. In addition, FISMA requires agencies to comply with National Institute of Standards and Technology (NIST) standards, and the Office of Management and Budget (OMB) requires agencies to comply with NIST guidelines.

---

[3] The *Federal Information Security Modernization Act of 2014* (FISMA 2014) (Pub. L. No. 113-283, Dec. 18, 2014) largely superseded the *Federal Information Security Management Act of 2002* (FISMA 2002), enacted as *Title III, E-Government Act of 2002*, Pub. L. No. 107-347, 116 Stat. 2899, 2946 (Dec. 17, 2002). As used in this report, FISMA refers both to FISMA 2014 and to those provisions of FISMA 2002 that were either incorporated into FISMA 2014 or were unchanged and continue in full force and effect.

NIST *Federal Information Processing Standards* (FIPS) Publication 199 requires agencies to categorize systems based on an assessment of the potential impact that a loss of confidentiality, integrity, or availability of such information or information system would have on organizational operations, organizational assets, individuals, other organizations, and the nation.[4] NIST FIPS 200[5] requires agencies to meet minimum security requirements by selecting the appropriate security controls, as described in NIST Special Publication (SP) 800-53.[6] This NIST publication provides a catalog of security and privacy controls for federal information systems and a process for selecting controls to protect organizational operations and assets.[7] The publication provides baseline security controls for low-, moderate-, and high-impact systems, and agencies have the ability to tailor or supplement their security requirements and policies based on agency mission, business requirements, and operating environment.

Further, in May 2017, the President issued an executive order[8] requiring agencies to immediately begin using NIST's Cybersecurity Framework[9] for managing their cybersecurity risks.

---

[4] The potential impact of a loss is categorized as one of three impact levels: 1) low – limited impact; 2) moderate – serious impact; and 3) high – severe or catastrophic impact. National Institute of Standards and Technology, *Standards for Security Categorization of Federal Information and Information Systems*, Federal Information Processing Standards Publication 199 (Gaithersburg, MD: February 2004).

[5] National Institute of Standards and Technology, *Minimum Security Requirements for Federal Information and Information Systems*, Federal Information Processing Standards Publication 200 (Gaithersburg, MD: March 2006).

[6] National Institute of Standards and Technology, *Security and Privacy Controls for Federal Information Systems and Organizations*, Special Publication 800-53, Revision 4 (Gaithersburg, MD: April 2013).

[7] Security control topics, referred to as families of security controls, covered by Special Publication 800-53 include access control, awareness and training, audit and accountability, security assessment and authorization, configuration management, contingency planning, identification and authentication, incident response, maintenance, media protection, physical and environmental protection, planning, personnel security, risk assessment, system and services acquisition, system and communications protection, system and information integrity, and program management.

[8] The White House, *Strengthening the Cybersecurity of Federal Networks and Critical Infrastructure*, Executive Order 13800 (Washington, D.C.: May 11, 2017).

[9] National Institute of Standards and Technology, *Framework for Improving Critical Infrastructure Cybersecurity* Version 1.1 (Gaithersburg, MD: Apr. 16, 2018). The framework was developed in response to an executive order issued by the prior administration, *Improving Critical Infrastructure Cybersecurity*, Executive Order 13636 (Washington, D.C.: Feb. 12, 2013). It was originally intended for use in protection of

The framework, which provides guidance for cybersecurity activities, is based on five core security functions:

- Identify: Develop the organizational understanding to manage cybersecurity risk to systems, assets, data, and capabilities.
- Protect: Develop and implement the appropriate safeguards to ensure delivery of critical infrastructure services.
- Detect: Develop and implement the appropriate activities to identify the occurrence of a cybersecurity event.[10]
- Respond: Develop and implement the appropriate activities to take action regarding a detected cybersecurity event.
- Recover: Develop and implement the appropriate activities to maintain plans for resilience and to restore any capabilities or services that were impaired due to a cybersecurity event.

According to NIST, these 5 functions occur concurrently and continuously, and provide a strategic view of the life cycle of an organization's management of cybersecurity risk. Within the 5 functions are 23 categories and 108 subcategories that include controls for achieving the intent of each function.[11] Appendix II provides a description of the framework categories and subcategories of controls.

---

critical infrastructure. NIST initially issued guidance in February 2014 and has since revised the framework.

[10] According to the National Institute of Standards and Technology, a cybersecurity event is defined as a cybersecurity change that may have an impact on organizational operations (including mission, capabilities, or reputation).

[11] For example, "risk assessment" is one of five categories that comprise the "identify" function. The risk assessment category is divided into six subcategories that involve activities such as identifying and documenting internal and external threats; identifying potential business impacts and likelihoods; and determining risk based on threats, vulnerabilities, likelihoods, and impacts. Each subcategory activity cross-references information system controls from various information security publications, including the National Institute of Standards and Technology's Special Publication 800-53.

Security Control Deficiencies Placed Selected CDC Systems at Risk

We reported in June 2018 that CDC had implemented numerous controls over the 24 systems we reviewed, but had not always effectively implemented controls to protect the confidentiality, integrity, and availability of these systems and the information maintained on them.

Table 2. Number of GAO-Identified Technical Control and Information Security Program Deficiencies at the Centers for Disease Control and Prevention and Associated Recommendations by Core Security Function

Core security function	Number of technical control deficiencies	Number of technical control recommendations	Number of information security program deficiencies	Number of information security program recommendations
Identify	0	0	5	5
Protect	85	161	1	1
Detect	8	18	3	3
Respond	1	5	1	1
Recover	0	0	1	1
Total	94	184	11	11

Source: GAO. | GAO-19-70.

Deficiencies existed in the technical controls[12] and agency-wide information security program[13] that were intended to (1) identify risk, (2) protect systems from threats and vulnerabilities, (3) detect cybersecurity events, (4) respond to these events, and (5) recover system operations. These deficiencies increased the risk that sensitive personally identifiable and health-related information, including information regarding the

[12] According to the National Institute of Standards and Technology, the technical controls (i.e., safeguards or countermeasures) for an information system are primarily implemented and executed by the information system through mechanisms contained in the hardware, software, or firmware components of the system. We also included physical security at data processing facilities as technical controls.

[13] The information security program includes processes, procedures, and practices used to manage the security of information systems.

transfer of biological agents and toxins dangerous to public health, could be disclosed or modified without authorization. As shown in table 2, deficiencies existed in all 5 core security function areas for the selected systems we reviewed.

CDC Had Identified Risk and Developed Policies and Plans, but Shortcomings Existed

Controls associated with the *identify* core security function are intended to help an agency develop an understanding of its resources and related cybersecurity risks to its systems, assets, data, and capabilities. These controls include identifying and assessing cybersecurity risk and establishing information security policies, procedures, and plans. We reported in June 2018 that, although CDC had taken steps to implement these controls, it had not (1) categorized the risk-related impact of a key system, identified threats, or reassessed risk for systems or facilities when needed; (2) sufficiently documented technical requirements in policies, procedures, and standards; and (3) described intended controls in facility security plans.

CDC Did Not Appropriately Categorize at Least One Key System, but Assessed Risk to Some Extent at System and Entity-Wide Levels

CDC Categorized Systems Based on Potential Impact of Compromise, but Did Not Appropriately Categorize a Key General Support System

As discussed earlier, FIPS Publication 199 requires agencies to categorize systems based on an assessment of the potential impact that a loss of confidentiality, integrity, or availability of such information or information system would have on organizational operations, organizational assets, individuals, other organizations, and the nation. For networks and other general support systems, NIST SP 800-60[14] notes that the categorization should be based on the high water mark[15] of supported information systems, and on the information types processed, transmitted across the network, or stored on the network or support system. Further, CDC's architecture design principles state that high-impact systems are to be maintained on dedicated machinery and be physically and logically secured from lower-risk systems.

CDC had categorized the 24 systems we reviewed, but the assigned impact level was not always appropriate. In this regard, the agency did not ensure that high-impact systems were logically secured from a lower-risk system. Specifically, seven selected high-impact systems relied on a general support system that the agency had categorized as a moderate-impact system (i.e., a lower-risk system). As a result, the high-impact systems were relying on controls in a less secure environment. Officials from the Office of the Chief Information Officer (OCIO) explained that the categorization of the supporting system was outdated based on changes to the agency's operating environment and that they planned to reevaluate the assigned impact level.

CDC Assessed Risk at the System Level, but Did Not Assess Threats, Document Risk-based Decisions, or Reassess Risk When Needed

According to NIST SP 800-30,[16] risk is determined by identifying potential threats to an organization and vulnerabilities in its systems, determining the likelihood that a particular threat may exploit vulnerabilities, and assessing the resulting impact on the organization's

---

[14] National Institute of Standards and Technology, Volume 1: *Guide for Mapping Types of Information and Information Systems to Security Categories*, Special Publication 800-60, Revision 1 (Gaithersburg, MD: August 2008).

[15] For an information system, the potential security impact levels assigned to each of the respective security objectives (confidentiality, integrity, availability) are the highest level (i.e., high water mark) for any one of these objectives that has been determined for the types of information resident on the information system. For example, if confidentiality is considered high impact, integrity moderate impact, and availability low impact, the system should be categorized as a high-impact system based on confidentiality being the highest level of the three objectives.

[16] National Institute of Standards and Technology, *Guide for Conducting Risk Assessments*, Special Publication 800-30, Revision 1 (Gaithersburg, MD: September 2012).

mission, including the effect on sensitive and critical systems and data. NIST also states that assessments should be monitored on an ongoing basis to keep current on risk-impacting changes to the operating environment.

CDC had developed system-level risk assessments for the 8 selected mission-essential systems, and had summarized its risks in a risk assessment report. However, only two of the eight risk assessments had identified potential threats, and only one of these assessments determined the likelihood and impact of threats to that system.

Further, CDC had not always documented risks associated with less secure configuration settings or monitored its assessments to address changes to the operating environment. For example, among the 94 technical control deficiencies that we identified for the 24 systems we reviewed, OCIO officials stated that the agency had not implemented controls for 20 deficiencies due to technical constraints.[17] However, CDC did not address risks associated with decisions not to implement controls for these reasons in the system risk assessments.

OCIO officials also partially attributed 5 of the 94 technical control deficiencies to new cybersecurity threats and to threat vectors that turned initially sound architecture decisions into vulnerabilities. However, CDC had not addressed such changes in the risk assessments for the affected systems. By not assessing threats or the likelihood of their occurrence and impact and by not documenting the risks, CDC cannot have assurance that appropriate controls are in place commensurate with the level of risk.

CDC Had a Process in Place to Assess Risk to Systems from an Entity-Wide Perspective

Beyond the system level, newly discovered threats or vulnerabilities may require an agency to make risk decisions from an entity-wide perspective. An entity-wide perspective is needed because the threats and vulnerabilities may affect more than specific systems.

CDC had a process in place to assess risk from an entity-wide perspective. This process included regular meetings among OCIO and

[17] A technical constraint could include, for example, implementing a control potentially causing functionality problems with legacy applications.

program office staff to discuss policy, threats, and incidents. Specifically, ISSOs held monthly meetings as a continuous monitoring working group to discuss policy updates. In addition, an OCIO official held quarterly briefings that included presentations on incident response tools, incident statistics, and potential threats. OCIO officials also held ad hoc meetings, as necessary, regarding vulnerability and threat concerns when the agency received email alerts from the Federal Bureau of Investigation, the Department of Homeland Security (DHS), or HHS.

CDC Had Not Updated Facility Risk Assessments

In addition to assessing risks for systems, agencies are to assess the risk to their facilities. The Interagency Security Committee (ISC)[18] requires agencies to determine the security level for federal facilities, and to conduct risk assessments at least once every 5 years for Level I and Level II facilities and at least once every 3 years for Level III, Level IV, and Level V facilities.[19]

However, the two facility risk assessments that we reviewed had not been updated in a timely manner.[20] Specifically, the risk assessments, covering Level III and Level IV facilities that house the 24 reviewed systems, had been last updated in January 2009 and March 2014—8 years earlier and just over 3 years earlier, at the time of our review in July 2017.

According to a CDC physical security official, the agency had previously relied on a third-party assessor to perform the assessments. The official also said that the agency planned to conduct its own facility risk

---

[18] The Interagency Security Committee, an interagency organization chaired by the Department of Homeland Security, was established by Executive Order No. 12977, 60 Fed. Reg. 54411 (October 1995), to enhance the quality and effectiveness of security and the protection of buildings and facilities in the United States occupied by federal employees for nonmilitary activities. Executive Order No. 12977 was later amended by Executive Order No. 13286, 68 Fed. Reg. 10619 (March 2003). The organization is comprised of senior level executives from federal agencies and departments.

[19] The facility security level is a categorization that ranges from Level I (lowest risk) to Level V (highest risk) based on the analysis of security-related factors, such as symbolism, population (e.g., employees and visitors), and size. The facility security level serves as the basis for identifying countermeasures that should be implemented at federal facilities to reduce the level of risk to an acceptable level.

[20] We reviewed two facility risk assessments because these two assessments covered facilities housing resources for the eight selected systems.

assessments and had recently developed procedures for conducting these assessments. Until it performs these assessments, CDC may not be aware of new risks to its facilities or the controls needed to mitigate the risks.

CDC Had Documented Controls in Policies, Procedures, and Standards, but Had Not Included Certain Technical Requirements

FISMA requires each agency to develop, document, and implement an information security program that, among other things, includes policies and procedures that (1) are based on a risk assessment, (2) cost-effectively reduce information security risks to an acceptable level, (3) ensure that information security is addressed throughout the life cycle of each system, and (4) ensure compliance with applicable requirements. According to NIST SP 800-53, an agency should develop policies and procedures for each of the 18 NIST families of security controls to facilitate the implementation of the controls.

CDC had documented numerous policies, procedures, and standards that addressed each of the 18 control families identified in NIST SP 800-53.

For example, the agency had developed policies and procedures governing physical access to CDC facilities, role-based training of personnel with significant security responsibilities, security assessment and authorization of systems, and continuity of operations, in addition to standard operating procedures that covered numerous other controls.

The agency had also developed the *CDC IT Security Program Implementation Standards*, which describes the agency's security program requirements and minimum mandatory standards for the implementation of information security and privacy controls. In addition, the agency had documented configuration standards, which specified minimum configuration settings, for devices such as firewalls, routers, switches, as well as Unix and Windows servers.

However, these policies and standards sometimes lacked the technical specificity needed to ensure controls were in place. To illustrate, the agency had not sufficiently documented detailed guidance or instructions to address numerous technical control deficiencies we identified, such as insecure network devices, insecure database configurations, not blocking certain email attachments, and not deploying a data loss prevention capability.

According to OCIO officials, the agency's periodic reviews and updates to existing cybersecurity policies and standards did not reveal and address these issues. Nevertheless, without clear and specific guidance or instructions for implementing technical controls, the agency had less assurance that controls were in place and operating as intended.

CDC Had Identified and Updated Controls in System Security Plans Annually, but Had Not Developed Facility Security Plans

FISMA requires each agency to develop, document, and implement an information security program that, among other things, includes subordinate plans for providing adequate information security for networks, facilities, and systems or a group of information systems, as appropriate. NIST states that plans should be reviewed and updated to ensure that they continue to reflect the correct information about the systems, such as changes in system owners, interconnections, and authorization status, among other things.

HHS and CDC policies require that such plans be reviewed annually. In addition, the ISC requires that agencies develop and implement an operable and effective facility security plan. CDC standards require the organization to prepare a facility security plan (or similar document).

CDC had developed security plans for the 8 selected mission-essential systems. With a few exceptions, the plans addressed the applicable security controls for those systems.[21] The agency also had reviewed and updated the plans annually.

However, CDC had not developed security plans for the facilities housing resources for the selected systems. Physical security officials stated that they had not developed security plans because they did not have a sufficient number of staff to develop them. Without comprehensive security plans for the facilities, CDC's information and systems would be at an increased risk that controls to address emergency situations would not be in place and personnel at the facilities would not be aware of their roles and responsibilities for implementing sound security practices to protect systems housed at these CDC locations.

CDC Had Implemented Controls Intended to Protect Its Systems, but Deficiencies Existed

The *protect* core security function is intended to help agencies develop and implement the appropriate safeguards for their systems to ensure achieving the agency's mission and to support the ability to limit or contain the impact of a potential cybersecurity event. Controls associated with this function include implementing controls to limit access to authorized users, processes or devices; encrypting data to protect its confidentiality and integrity; configuring devices securely and updating software to protect systems from known vulnerabilities; and providing training for cybersecurity awareness and performing security-related duties. Although CDC had implemented controls that were intended to protect its operating environment, we reported in June 2018 that the agency did not consistently (1) implement access controls effectively, (2) encrypt sensitive data, (3) configure devices securely or apply patches in a timely manner, or (4) ensure staff with significant security responsibilities received role-based training.

[21] We determined that, of the approximately 2,800 total controls listed in the security plans for the 8 selected mission-essential systems, CDC had not provided detailed implementation information for 57 of the controls (or about 2 percent).

CDC Did Not Consistently Implement Effective Access Controls

A basic management objective for any agency is to protect the resources that support its critical operations from unauthorized access. Agencies accomplish this objective by designing and implementing controls that are intended to prevent, limit, and detect unauthorized access to computing resources, programs, information, and facilities. Access controls include those related to identifying and authenticating users, authorizing access needed to perform job duties, protecting system boundaries, and physically protecting information system assets. However, CDC had not consistently implemented these controls.

CDC Implemented Enterprise-Wide Identification and Authentication Controls, but Did Not Consistently and Securely Configure Password Controls for Certain Accounts on Devices and Systems

NIST SP 800-53 states that agencies should implement multi-factor authentication for their users of information systems. Multi-factor authentication involves using two or more factors to achieve authentication. A factor is something you know (password or personal identification number), something you have (token and personal identity verification (PIV) card), or something you are (biometric). Also, NIST and CDC policy state that information systems shall have password management controls established to include minimum password complexity requirements, password lifetime restrictions, prohibitions on password reuse, and user accounts temporarily locked out after a certain number of failed login attempts during a specified period of time.

CDC had applied enterprise-wide solutions to ensure appropriate identification and multi-factor authentication of its general user community through, for example, the use of PIV cards. However, instances of weak password management controls existed for certain accounts on network devices, servers, and database systems. According to OCIO officials, password control deficiencies existed primarily due to technical constraints, administrators not being aware of technical requirements, or administrators not adequately monitoring configuration settings. Without more secure password settings, CDC's information and systems are at an increased risk that unauthorized individuals could have guessed passwords and used them to obtain unauthorized access to agency systems and databases.

CDC Authorized Users More Access than Needed to Perform Their Jobs

NIST SP 800-53 states that agencies should employ the principle of least privilege, allowing only authorized access for users (or processes acting on behalf of users) that are necessary to accomplish assigned tasks. It also states that privileged accounts—those with elevated access permissions—should be strictly controlled and used only for their intended administrative purposes.

CDC had implemented controls intended to ensure that users were granted the minimum level of access permissions necessary to perform their legitimate job-related functions. However, the agency had granted certain users more access than needed for their job functions, including excessive access permissions on a key server.

According to OCIO officials, CDC systems had deficiencies related to restricting access primarily due to technical constraints or administrators not adequately monitoring configuration settings. By not appropriately restricting access, CDC's information and systems are at an increased risk that individuals could deliberately or inadvertently compromise database systems or gain inappropriate access to information resources.

CDC Did Not Effectively Implement Boundary Controls to Ensure Network Integrity

NIST SP 800-53 states that agencies should control communications at information systems' external boundaries. It states that, to manage risks, agencies should use boundary protection mechanisms to separate or partition computing systems and network infrastructures containing higher-risk systems from lower-risk systems.[22]

---

[22] Boundary protection controls logical connectivity into and out of networks and controls connectivity to and from devices connected to a network. Implementing multiple layers of security to protect an information system's boundaries can reduce the risk of a successful

Although CDC had implemented multiple controls that were designed to protect system boundaries, the agency had not sufficiently separated higher-risk systems from lower-risk systems. According to OCIO officials, deficiencies in boundary protection controls existed due to new cybersecurity threats turning initially sound architecture decisions into vulnerabilities, technical constraints, and administrators not being aware of technical requirements or adequately monitoring configuration settings. Without stronger boundary controls, CDC's information and systems are at an increased risk that an attacker could have exploited these boundary deficiencies and leveraged them to compromise CDC's internal network.

CDC Physically Protected Information System Assets, but Did Not Consistently Ensure Access Remained Appropriate

NIST SP 800-53 states that agencies should implement physical access controls to protect employees and visitors, information systems, and the facilities in which they are located. In addition, NIST states that agencies should review access lists detailing authorized facility access by individuals at the agency-defined frequency. In its standards, CDC requires implementation of the NIST special publication and requires that access lists detailing authorized facility access by individuals be reviewed at least every 365 days.

CDC had implemented physical security controls. The agency had implemented physical security measures to control access to certain areas and to ensure the safety and security of its employees, contractors, and visitors to CDC facilities. For example, CDC had issued PIV cards and Cardkey Proximity Cards to its employees and contractors, and had limited physical access to restricted areas based on the permissions it granted via these cards.

However, the agency had not consistently reviewed authorized access lists. In this regard, CDC did not have a process in place for periodically reviewing the lists of individuals with access to rooms containing sensitive

cyberattack. For example, multiple firewalls can be deployed to prevent both outsiders and trusted insiders from gaining unauthorized access to systems, and intrusion detection technologies can be deployed to defend against attacks from the Internet.

resources to ensure that such access remained appropriate. Without reviewing authorized access lists, CDC has reduced assurance that individual access to its computing resources and sensitive information is appropriate.

CDC Had Not Consistently Encrypted Sensitive Authentication Data

NIST SP 800-53 states that agencies should encrypt passwords both while stored and transmitted, and configure information systems to establish a trusted communication path between the user and the system.

Additionally, NIST requires that, when agencies use encryption, they use an encryption algorithm that complies with FIPS 140-2.[23]

CDC had used FIPS-compliant encryption for its PIV card implementation, but had not effectively implemented encryption controls in other areas. According to OCIO officials, encryption control deficiencies existed primarily due to technical constraints, administrators not being aware of a technical solution, or configuration settings not being adequately monitored. By not using encryption effectively, CDC limits its ability to protect the confidentiality of sensitive information, such as passwords.

CDC Had Not Consistently Configured Servers Securely or Applied Patches in a Timely Manner

NIST SP 800-53 states that agencies should disable certain services with known security vulnerabilities. This includes configuring security control settings on operating systems in accordance with publicly available security checklists (or benchmarks) promulgated by NIST's National Checklist Program repository. This repository contains, for example, the security configuration benchmarks established by the Center for Internet Security (CIS) for Windows servers.

NIST also states that agencies should test and install newly-released security patches, service packs, and hot fixes in a timely manner. In

---

[23] National Institute of Standards and Technology, *Security Requirements for Cryptographic Modules*, Federal Information Processing Standards Publication 140-2 (Gaithersburg, MD: May 25, 2001).

addition, CDC policy required that software patches for remediating vulnerabilities designated as critical or high risk be applied to servers within 45 days of being notified that a patch is available or within 7 days of when an exploit is known to exist. Further, agency policy specified that administrators configure Windows servers in accordance with the CDC-approved security benchmarks.

CDC had documented security configuration baselines, but had not always securely configured its systems or applied patches. In addition, the agency had not consistently configured security settings in accordance with prescribed security benchmarks or applied patches in a timely manner. For example:

- CDC had configured Windows servers to run unnecessary services.
- CDC had configured only about 62 percent of the security settings in accordance with prescribed benchmark criteria on the Windows and infrastructure servers supporting five systems that we reviewed.
- During our site visit in April 2017, CDC had not installed 21 updates on about 20 percent of the network devices, including 17 updates that the vendor considered to be critical or high-risk. The oldest of the missing updates dated back to January 2015.
- CDC had not updated database software supporting two selected systems to a more recent version that addressed vulnerabilities with a medium severity rating.

According to OCIO officials, CDC had deficiencies in configuration and patching primarily due to administrators not being aware that there was a technical solution or did not adequately monitor configuration settings. By not securely configuring devices and installing updates and patches in a timely manner, the agency is at increased risk that individuals could have exploited known vulnerabilities to gain unauthorized access to agency computing resources.

Staff Received Security Awareness Training, but at Least 15 Percent of Those with Significant Security Responsibilities Did Not Receive Role-Based Training

According to NIST SP 800-53, agencies should provide adequate security training to individuals in a role such as system/network administrator and to personnel conducting configuration management and auditing activities, tailoring the training to their specific roles. In addition, one of the cybersecurity cross-agency priority goals requires that agencies implement training that reduces the risk that individuals will introduce malware through email and malicious or compromised web sites.[24]

Consistent with NIST SP 800-53, CDC policy required network users to receive annual security awareness training. Accordingly, for fiscal year 2017, all CDC staff completed the required annual security awareness training.

CDC policy also required that those staff identified as having significant security responsibilities receive role-based training every 3 years. However, not all staff with significant security responsibilities received role-based training within the defined time frames. The agency used a tracking system to monitor the status of role-based training for 377 individuals who had been identified as having significant security responsibilities. As of May 2017, 56 (about 15 percent) of the 377 individuals had not completed the training within the last 3 years, and 246 (about 65 percent) of them had not taken training within the last year.

In addition, CDC had not identified at least 30 other staff with significant security responsibilities who required role-based training. Specifically, none of the 18 security and database administrators for four selected systems were included among the individuals being tracked, although these administrators had significant security responsibilities. Further, the agency provided us with a list of 42 individuals whose job series indicated that they required role-based training. However, 12 of the

---

[24] Cybersecurity cross-agency priority goals were established by the prior administration as part of implementing the requirement in the *Government Performance and Results Act Modernization Act of 2010* to develop federal government priority goals for information technology management. Sec. 5, Pub. L. No. 111-352 (Jan. 4, 2011); 124 Stat. 3866, 3873; 31 U.S.C. § 1120(a)(1)(B).

42 were not included among the tracked individuals. Furthermore, given the number of deficiencies identified and the rapidly evolving nature of cyber threats, CDC's requirement that staff take role-based training only once every 3 years is not sufficient for individuals with significant cybersecurity responsibilities.

According to OCIO officials, managers are responsible for identifying those individuals with significant security responsibilities. The process used to track training was manual and required an individual's manager to specify training requirements. The officials noted that the agency plans to implement a new HHS annual role-based training requirement in fiscal year 2018 and that they intend to work to enhance oversight as the new requirement is implemented.[25]

The officials also stated that at least 10 of the 94 technical control-related deficiencies identified in our June 2018 report had resulted, at least in part, from staff not being aware of control requirements or solutions to address the deficiencies. As a result, CDC's information and systems are at increased risk that staff may not have the knowledge or skills needed to appropriately protect them.

CDC Had Not Effectively Implemented Controls Intended to Detect Incidents or Deficiencies

The *detect* core security function is intended to allow for the timely discovery of cybersecurity events. Controls associated with this function include logging and monitoring system activities and configurations, assessing security controls in place, and implementing continuous monitoring. In June 2018, we reported that, although CDC had implemented controls intended to detect the occurrence of a cybersecurity event, it had not sufficiently implemented logging and monitoring capabilities or effectively assessed security controls.

---

[25] Since the issuance of the June 2018 report, CDC has updated their role-based training requirement from once every three years to annually.

CDC Had Implemented Limited Logging and Monitoring Capabilities

NIST SP 800-53 states that agencies should enable system logging features and retain sufficient audit logs to support the investigations of security incidents and the monitoring of select activities for significant security-related events. In addition, National Archives and Records Administration records retention guidance states that system files containing information requiring special accountability[26] that may be needed for audit or investigative purposes should be retained for 6 years after user accounts have been terminated or passwords altered, or when an account is no longer needed for investigative or security purposes, whichever is later.[27] NIST also states that agencies should monitor physical access to facilities where their information systems reside to detect physical security incidents. Further, NIST SP 800-53 states that agencies should monitor and control changes to configuration settings.

Although CDC had implemented centralized logging and network traffic monitoring capabilities, the capabilities were limited. For example, the agency's centralized logging system used for security monitoring had a limited storage capacity and did not meet the National Archives and Records Administration requirements. In addition, CDC had not centrally collected and monitored security event data for many key assets connected to the network. As a result, increased risk existed that CDC would not have been able to detect anomalous activities that may have occurred from malware attacks over time. OCIO officials stated that, as a compensating measure, the agency prevents direct communications between workstations. However, such a measure does not allow the agency to detect potentially inconsistent activities that may have occurred from malware attacks within the same data center.

CDC also had not consistently reviewed physical access logs to detect suspicious physical access activities, such as access outside of normal work hours and repeated access to areas not normally accessed. Program

---

[26] Files containing information requiring special accountability are user identification records associated with systems which are highly sensitive and potentially vulnerable.

[27] National Archives and Records Administration, *General Records Schedule 3.2: Information Systems Security Records*, Transmittal 26 (Washington D.C.: September 2016).

offices responsible for 7 of the 8 selected mission-essential systems did not conduct such a review. According to OCIO officials, the offices were not aware of the need for a review. However, without reviewing physical access logs, CDC has reduced assurance that the agency would detect suspicious physical access activities.

Further, CDC had not routinely monitored the configuration settings of its systems to ensure that the configurations were securely set. For example, for at least 41 of 94 technical control deficiencies we identified, OCIO officials cited quality control gaps where the change management process or system administrators had not discovered deficiencies resulting from insecure configuration settings. Without an effective monitoring process in place for system configurations, the agency was not aware of insecure system configurations.

CDC Did Not Effectively Test or Assess Controls to Detect Deficiencies

FISMA requires each agency to periodically test and evaluate the effectiveness of its information security policies, procedures, and practices. The law also requires agencies to test the management, operational, and technical controls for every system identified in the agency's required inventory of major information systems at a frequency depending on risk, but no less than annually. In addition, NIST SP 800- 53A identifies three assessment methods—interview, examine, and test—and describes the potential depth and coverage for each.[28] Assessing a control's effectiveness based on an interview is likely less rigorous than examining a control; similarly, examining a control is likely less rigorous than testing the control's functionality.

---

[28] According to NIST, the interview method is the process of holding discussions with individuals or groups of individuals within an organization to facilitate assessor understanding, achieve clarification, or obtain evidence. The examine method is the process of reviewing, inspecting, observing, studying, or analyzing one or more assessment objects (i.e., specifications, mechanisms, or activities) to facilitate assessor understanding, achieve clarification, or obtain evidence. The test method is the process of exercising one or more assessment objects (i.e., activities or mechanisms) under specified conditions to compare actual with expected behavior.

CDC had not sufficiently tested or assessed the effectiveness of the security controls for the 8 mission-essential systems that we reviewed. Although CDC annually assessed security controls of selected systems, the agency had only examined control descriptions in security plans to ensure accuracy. At least once every 3 years, the agency selected controls for a more in-depth assessment of the 8 mission-essential systems we reviewed. However, CDC had assessed only 191 (about 7 percent) of 2,818 controls described in the security plans for the selected systems. In addition, the agency used methods for assessing controls that were often not rigorous enough to identify the control deficiencies that we identified. For example, as depicted in Figure 1, CDC relied exclusively on interviews—a less rigorous method—to assess 20 percent of the 191 controls it assessed for the selected systems.

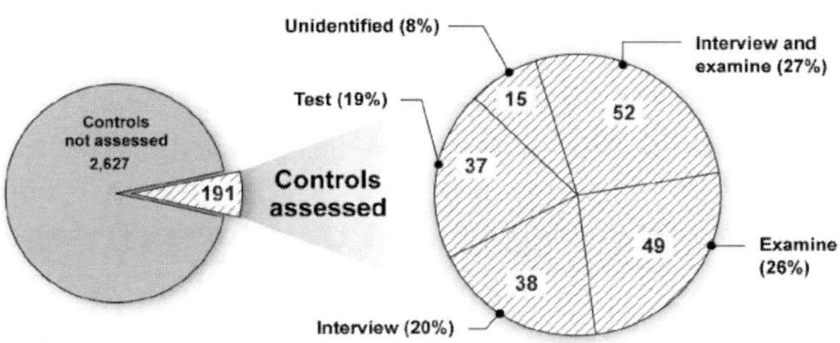

Source: GAO analysis of selected system assessments. | GAO-19-70.

Figure 1. Methods Used by the Centers for Disease Control and Prevention to Assess Security Controls.

The security control tests and assessments were insufficient in part because CDC had not developed comprehensive security assessment plans or had not consistently implemented the plans for the 8 selected mission-essential systems we reviewed. For example, one system's assessment plan indicated that five controls should be assessed using a testing methodology; instead, however, the assessor conducted interviews to determine whether controls were effective or not.

OCIO officials stated that the security control test and assessment process is manual and staffing is limited. They stated that the agency intends to rely increasingly on automated tools—such as the tools implemented by the Continuous Diagnostics and Mitigation program—for performing the assessments. Nevertheless, by not assessing controls in an in-depth and comprehensive manner, CDC has limited assurance that the security controls are in place and operating as intended. Further, without developing and implementing comprehensive assessment plans, assessments may not be performed with sufficient rigor to identify control deficiencies.

CDC Had Implemented Processes for Responding to Incidents or Identified Deficiencies, but Did Not Always Take Timely Corrective Actions

The *respond* core security function is intended to support the ability to contain the impact of a potential cybersecurity event. Controls associated with this function include implementing an incident response capability and remediating newly-identified deficiencies. Although CDC had implemented controls for incident response to detect cybersecurity events, we reported in June 2018 that the agency had not maintained adequate information to support its incident response capability or taken timely corrective actions to remediate identified control deficiencies.

CDC Had Implemented Incident Response Capabilities, but Did Not Maintain Adequate Information

NIST SP 800-53 and SP 800-61 state that agencies should develop and document an incident response policy with corresponding implementation procedures and an incident response plan, and keep them updated according to agency requirements. NIST also states that agencies should implement an incident handling capability, including an incident response team that consists of forensic/malicious code analysts. In addition, agencies are to provide incident response training for the team and test the incident response capability to determine the effectiveness of the response.

Further, NIST states that agencies are to monitor incidents by tracking and documenting them and maintain records about each incident, including forensic analysis. Finally, National Archives and Records Administration guidance states that records and data relevant to security incident investigations should be retained for 3 years.

CDC had implemented an incident response capability. The agency had developed policy, procedures, and a plan that addressed incident response, and updated them annually. CDC had an incident response team that managed all of the incident handling and response efforts for the agency, and conducted forensic analyses for reported security incidents. Team members had undergone training, such as an advanced network forensic and analysis course offered by a private firm. In addition, the agency had periodically tested its incident handling capability by conducting penetration testing exercises. These exercises allowed the team to test its real-time response capabilities.

CDC's incident response procedures state that incident tickets should include a description of actions taken, response time, and whether actions have been completed or not. The agency's procedures also require that computers affected by an incident be removed from the network immediately.

Nevertheless, CDC had shortcomings in implementing its incident response capability and monitoring procedures. For the 11 security incidents CDC considered most significant over a 19-month period ending in March 2017,[29] the agency had not consistently described the actions taken, the response times, or whether remedial actions had been completed. The agency also had not maintained audit log records for its security incidents. For example, the agency described recommended actions for 10 of the 11 incidents, but did not describe the actions that had been taken.

In addition, although incident response team officials told us that all incident ticket records had been saved, CDC had not retained system log data that supported incident resolution for at least five of the incidents. The agency's policy did not address record retention in accordance with

[29] We conducted our initial site visit to assess incident response capabilities during the week of March 7, 2017.

National Archives and Records Administration guidance. Further, for two of the security incidents, the security incident tickets did not clearly indicate when two compromised workstations had been removed from the network. According to OCIO officials, shortcomings in fully documenting incidents resulted from the organization being understaffed, primarily due to budget limitations and the inability to hire qualified personnel. Without effectively tracking and documenting information system security incidents, CDC's systems are at increased risk that the impact of security incidents would not be fully addressed.

CDC Had Remedial Action Plans to Address Identified Deficiencies for Selected Systems, but Did Not Always Take Timely Corrective Actions or Have Plans for Other Needed Corrective Actions

FISMA requires each agency to develop, document, and implement an information security program that, among other things, includes a process for planning, implementing, evaluating, and documenting remedial actions to address any deficiencies in information security policies, procedures, or practices. NIST SP 800-53 states that agencies are to develop a plan of action and milestones (POA&M) for an information system to document the agency's planned remedial actions to correct identified deficiencies. CDC policy was consistent with the NIST guidelines.

CDC had developed POA&Ms for deficiencies identified by its security control assessments, but had not remediated the deficiencies in a timely manner. For each of the 8 selected mission-essential systems, the agency had created plans for correcting control deficiencies. However, the agency did not implement several remedial actions by their due date. For example, expected completion dates had passed for correcting deficiencies associated with 4 of the 8 selected mission-essential systems. For these 4 systems, the completion dates were 1 to 8 months beyond the due dates at the time of our review in September 2017.

According to Office of the Chief Information Security Officer officials, program offices that own the systems did not always communicate updates on the status of remedial actions for their respective systems, noting that deficiencies may have been corrected. Without effective communication to update its POA&Ms, CDC was not in a position to effectively manage its remedial actions and correct known deficiencies in a timely manner.

CDC Had Developed and Tested Plans for System Recovery, but Had Not Assessed the Risk Associated with the Close Proximity of an Alternate Processing Site

The *recover* core security function is intended to support timely recovery of normal operations to reduce the impact from a cybersecurity event. Controls associated with this function include developing and testing contingency plans to ensure that, when unexpected events occur, critical operations can continue without interruption or can be promptly resumed, and that information resources are protected. Losing the capability to process, retrieve, and protect electronically maintained information can significantly affect an agency's ability to accomplish its mission. If contingency planning is inadequate, even relatively minor interruptions can result in lost or incorrectly processed data, which can cause financial losses, expensive recovery efforts, and inaccurate or incomplete information.

NIST SP 800-53 states that agency systems should have a contingency plan that includes the identification of key personnel and the systems' essential mission functions and addresses full information system restoration. For high-impact systems, NIST specifies that agencies test contingency plans at an alternate processing site that is separated from the primary processing site to reduce susceptibility to the same threats. In addition, NIST states that organizations should initiate corrective actions based on testing if they are needed.

As we reported in June 2018, CDC had developed and fully tested contingency plans for each of the 8 selected mission-essential systems that we reviewed.

Each plan identified key personnel and their contact information, essential mission functions of the systems, and instructions on how to fully restore the systems in the event of a disruption. Additionally, between January 2015 and May 2017, CDC had tested whether the 8 systems could be recovered at their respective alternate sites, and had initiated corrective actions based on the results of the tests.

However, the alternate site for 6 of the 8 selected mission-essential systems was located in relatively close proximity to the main processing site. Although 2 systems had alternate sites located in another state, the alternate site for the other 6 systems was within the same metropolitan area. As a result, an event such as a natural disaster or substantial power outage could affect both the main and alternate sites for these systems, potentially rendering CDC unable to complete functions associated with its mission. Prompt restoration of service is necessary because the required recovery time for these systems ranged from 4 to 24 hours.

Security plans for 3 of the systems recognized the hazards of having the sites within the same geographical region, but stated that CDC had accepted this risk. According to OCIO officials, having a site further away was cost prohibitive; however, the officials had not documented this analysis or the associated risk of having the agency's processing sites located within the same geographical area. Without documenting the analysis and associated risk, CDC had less assurance that senior leadership was aware of the risk of agency systems being unavailable. As a consequence, senior leadership may not agree whether acceptance of the risk was warranted.

CDC Had Not Consistently or Effectively Implemented Elements of Its Information Security Program

An underlying reason for the information security deficiencies in selected systems was that, although the agency had developed and documented an agency-wide information security program, it had not consistently or effectively implemented elements of the program.

FISMA requires each agency to develop, document, and implement an information security program that, among other things, includes the following elements:

- periodic assessments of the risk and magnitude of the harm that could result from the unauthorized access, use, disclosure, disruption, modification, or destruction of information and information systems that support the operations and assets of the agency;
- policies and procedures that (1) are based on risk assessments, (2) cost-effectively reduce information security risks to an acceptable level, (3) ensure that information security is addressed throughout the life cycle of each system, and (4) ensure compliance with applicable requirements;
- plans for providing adequate information security for networks, facilities, and systems or group of information systems, as appropriate;
- security awareness training to inform personnel of information security risks and of their responsibilities in complying with agency policies and procedures, as well as training personnel with significant security responsibilities for information security;
- periodic testing and evaluation of the effectiveness of information security policies, procedures, and practices, to be performed with a frequency depending on risk, but no less than annually, and that includes testing of management, operational, and technical controls for every system identified in the agency's required inventory of major information systems;
- a process for planning, implementing, evaluating, and documenting remedial actions to address any deficiencies in the information security policies, procedures, or practices of the agency; and

- plans and procedures to ensure continuity of operations for information systems.

As discussed previously in this chapter, CDC had implemented aspects of each of these elements. For example, the agency had conducted risk assessments, developed security plans, assessed security controls, developed remedial action plans, and developed and tested contingency plans for each of the 8 selected mission-essential systems. In addition, the agency had documented numerous policies and procedures and ensured that staff had completed annual security awareness training.

However, CDC's program had shortcomings. For example, as discussed earlier in this chapter, CDC had not consistently or effectively:

- addressed threats, technical constraints, and the changing threat environment in its system risk assessments, or assessed the risk of having alternate processing sites within close proximity to each other;
- documented detailed technical requirements in policies and procedures, or facility controls in facility security plans;
- tracked and trained staff with significant security responsibilities;
- monitored configuration settings and comprehensively assessed system controls;
- remediated deficiencies in a timely manner; or
- documented its cost analysis and associated risk of having an alternate processing site within the same geographical region as its primary processing site.

Until CDC addresses these shortcomings and consistently and effectively implements all elements of its information security program, the agency will lack reasonable assurance that its computing resources are protected from inadvertent or deliberate misuse.

CDC Has Implemented Many of the Recommendations in Our June 2018 Report and Plans to Implement the Rest

In our June 2018 report,[30] we made 195 recommendations to CDC to strengthen its technical security controls and bolster its agency-wide information security program. Specifically, we recommended that the agency take 184 actions to resolve technical control deficiencies by implementing stronger access controls, encrypting sensitive data, configuring devices securely, applying patches in a timely manner, strengthening firewall rules, and implementing logging and monitoring controls more effectively, among other actions. We also made 11 recommendations for CDC to improve its information security program by, among other things, assessing risks as needed, documenting more detailed technical requirements, monitoring and assessing controls more comprehensively, and remediating deficiencies in a timely manner.

Since the issuance of our June 2018 report, CDC has made significant progress in implementing the recommendations we made to resolve the technical security control deficiencies in the information systems we reviewed and to improve its information security program. In this regard, the agency has implemented many of the recommendations for improving technical security controls for the systems we reviewed and has developed plans to implement recommendations for enhancing its information security program.

Specifically, as of August 3, 2018, CDC had fully implemented 102 (55 percent) of the 184 recommendations we made to fortify the technical security controls over the systems we reviewed. In addition, the agency had partially implemented 20 (11 percent) of the 184 recommendations. In these instances, CDC had made progress toward implementing the recommendations, but had not completed all of the necessary corrective actions for us to close the recommendations. Therefore, these recommendations remain open. Further, CDC did not provide any evidence

---

[30] GAO-18-437SU.

that it had implemented the remaining 62 technical control-related recommendations.

Table 3 summarizes the status of CDC's efforts to implement the 184 recommendations that we made to resolve the technical control deficiencies, as of August 3, 2018.[31]

Table 3. Status of Efforts by the Centers for Disease Control and Prevention to Implement GAO's Technical Control-Related Recommendations by Core Security Function, as of August 3, 2018

Core security function	Number of technical control-related recommendations	Status of Technical Control-related Recommendations		
n/a	n/a	Closed–implemented	Open–partially implemented	Open–not implemented
Identify	0	0	0	0
Protect	161	89	18	54
Detect	18	9	2	7
Respond	5	4	0	1
Recover	0	0	0	0
Totals	184	102	20	62

Legend:
Closed-implemented (CDC successfully completed actions to implement the recommendation)
Open-partially implemented (CDC had made progress toward—but has not completed—implementing the recommendation) Open-not implemented (CDC has not acted to implement the recommendation)
Source: GAO analysis of CDC data. | GAO-19-70.

By implementing 102 recommendations, CDC (as of August 3, 2018) reduced some of the risks associated with certain key activities. Specifically, these efforts included protecting network boundaries and logging and monitoring security events for indications of inappropriate or unusual activity on systems—that we highlighted in our June 2018 report as being particularly vulnerable and requiring the agency's greater priority

[31] We determined that 14 of the implemented, and 9 of the partially-implemented recommendations were either fully or partially addressed due to the decommissioning of one of the mission-essential systems, a key application process, and certain servers selected for this review. In August 2018, while onsite, we reviewed configuration settings—to include the authorization to operate, and change management procedures— to ensure the vulnerabilities were addressed in the new systems and servers.

and attention. In addition, the agency had implemented several of our recommendations to rectify a number of the security control deficiencies. These efforts included strengthening firewall rules, implementing stronger access controls, configuring devices securely, and expanding its audit monitoring capabilities.

In addition, CDC had developed a plan of action and milestones (POA&M) for each of the identified technical control deficiencies and related recommendations that remained open as of August 3, 2018. The POA&Ms assigned organization responsibilities, identified estimated costs, identified points of contact, and established time frames for resolving the deficiencies and closing the related recommendations. The agency's plans called for it to implement the majority of the remaining open technical control-related recommendations by September 2019, and all recommendations by September 2020, as shown in Figure 2.

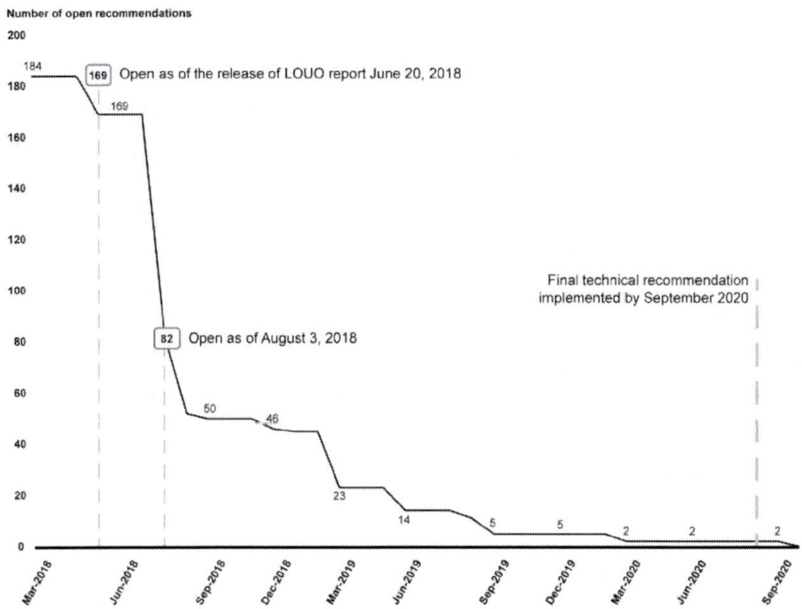

Source: GAO analysis of Centers for Disease Control and Prevention data. | GAO-19-70.

Figure 2. The Centers for Disease Control and Prevention's Planned Timeline for Fully Implementing GAO's Technical Control-Related Recommendations.

Our June 2018 report also included 11 recommendations to CDC to improve its information security program. In particular, we recommended that the agency, among other things, evaluate system impact level categorizations to ensure they reflect the current operating environment; update risk assessments to identify threats and the likelihood of impact of the threat on the environment; and update the facility risk assessments. In addition, we recommended that the agency take the necessary steps to make sure staff with significant security roles and responsibilities are appropriately identified and receive role-based training; monitor the configuration settings of agency systems to ensure the settings are set as intended; update security control assessments to include an assessment of controls using an appropriate level of rigor; and remediate POA&Ms in a timely manner. Further, we recommended that the agency document the cost-benefit analysis with associated risk of having an alternate site within the same geographical region as the main site.

As of August 3, 2018, the agency had partially implemented 1 of the 11 information security program-related recommendations, but had not provided any evidence that it had implemented the remaining 10 recommendations. Regarding the partially implemented recommendation, CDC had provided role-based training to all personnel performing significant security responsibilities. However, the agency still needed to establish and automate the identification process and the tracking of training records for individuals needing specialized security role-based training. CDC had developed plans to fully implement this recommendation and each of the remaining 10 information security program-related recommendations by July 2019. Fully implementing the open recommendations is essential to ensuring that the agency's systems and sensitive information are not at increased and unnecessary risk of unauthorized use, disclosure, modification, or disruption.

AGENCY COMMENTS

We received written comments on a draft of this chapter from CDC. In its comments, which are reprinted in appendix III, the agency stated that it recognizes the risks associated with operating a large, global information technology enterprise and has implemented processes, procedures, and tools to better ensure the prevention, detection, and correction of potential incidents. CDC also said cybersecurity remains a high priority and that it takes the responsibilities for protecting public health information and data entrusted to it seriously. To strengthen its cybersecurity program, the agency stated that it is restructuring and streamlining the cyber program and IT infrastructure of its Office of the Chief Information Officer.

Further, CDC stated that it has leveraged GAO's limited official use only report, issued in June 2018, to accelerate its implementation, infrastructure, and software deployments to complete phrases one and two of DHS's Continuous Diagnostics and Mitigation program. The agency also said it concurred with, and highlighted a number of actions that it had planned or begun taking to remediate, the 11 security program recommendations that we made to CDC in our June 2018 report.

We are sending copies of this chapter to the appropriate congressional committees, the Secretary of Health and Human Services, and the department's Office of the Inspector General, the Director of CDC, and interested congressional parties. In addition, the report will be available at no charge on the GAO website at http://www.gao.gov.

Gregory C. Wilshusen
Director, Information Security Issues

Nabajyoti Barkakati
Chief Technologist

APPENDIX I: OBJECTIVES, SCOPE, AND METHODOLOGY

Our objective was to assess the extent to which CDC had effectively implemented an information security program and controls to protect the confidentiality, integrity, and availability of its information on selected information systems. In June 2018, we issued a report which detailed the findings from our work in response to this objective.[32] In the report, we made 184 recommendations to CDC to resolve the technical security control deficiencies in the information systems we reviewed and 11 additional recommendations to improve its information security program. We designated that report as "limited official use only" (LOUO) and did not release it to the general public because of the sensitive information it contained.

This chapter publishes the findings discussed in our June 2018 report, but we have removed all references to the sensitive information. Specifically, we deleted the names of the information systems and computer networks that we examined, disassociated identified control deficiencies from named systems, deleted certain details about information security controls and control deficiencies, and omitted an appendix that was contained in the LOUO report. The appendix contained sensitive details about the technical security control deficiencies in the CDC's information systems and computer networks that we reviewed, and the 184 recommendations we made to mitigate those deficiencies. We also provided a draft of this chapter to CDC officials to review and comment on the sensitivity of the information contained herein and to affirm that the report can be made available to the public without jeopardizing the security of CDC's information systems and networks.

In addition, this chapter addresses a second objective that was not included in the June 2018 report. Specifically, this objective was to determine the extent to which CDC had taken corrective actions to address the previously identified security program and technical control

[32] GAO, *Information Security: CDC Needs to Improve Its Program and Resolve Control Deficiencies*, GAO-18-437SU (Washington, D.C.: June 20, 2018).

deficiencies and related recommendations for improvement that we identified in the earlier report.

As noted in our June 2018 report, we determined the extent to which CDC had effectively implemented an information security program and controls to protect the confidentiality, integrity, and availability of its information on selected information systems. To do this, we initially gained an understanding of the overall network environment, identified interconnectivity and control points, and examined controls for the agency's networks and facilities. We conducted site visits at two CDC facilities in Atlanta, Georgia.

To evaluate CDC's controls over its information systems, we used our Federal Information System Controls Audit Manual,[33] which contains guidance for reviewing information system controls that affect the confidentiality, integrity, and availability of computerized information. We based our assessment of controls on requirements identified by the Federal Information Security Modernization Act of 2014 (FISMA),[34] which establishes key elements for an effective agency-wide information security program; NIST guidelines and standards;[35] Department of Health and Human Services and CDC policies, procedures, and standards; and standards and guidelines from relevant security organizations, such as the

---

[33] GAO, *Federal Information System Controls Audit Manual* (FISCAM), GAO-09-232G (Washington, D.C.: February 2009).

[34] *The Federal Information Security Modernization Act of 2014* (FISMA 2014) (Pub. L. No. 113-283, Dec. 18, 2014) largely superseded the *Federal Information Security Management Act of 2002* (FISMA 2002), enacted as Title III, *E-Government Act of 2002*, Pub. L. No. 107-347, 116 Stat. 2899, 2946 (Dec. 17, 2002). As used in this report, FISMA refers both to FISMA 2014 and to those provisions of FISMA 2002 that were either incorporated into FISMA 2014 or were unchanged and continue in full force and effect.

[35] For example, see National Institute of Standards and Technology, *Minimum Security Requirements for Federal Information and Information Systems*, Federal Information Processing Standards Publication 200 (Gaithersburg, MD: March 2006), and National Institute of Standards and Technology, *Security and Privacy Controls for Federal Information Systems and Organizations*, Special Publication 800-53, Revision 4 (Gaithersburg, MD: April 2013).

National Security Agency, the Center for Internet Security,[36] and the Interagency Security Committee.[37]

We had reviewed a non-generalizable sample of the agency's information systems, focusing on those systems that (1) collect, process, and maintain private or potentially-sensitive proprietary business, medical, and personally identifiable information; (2) are essential to CDC's mission; and (3) were assigned a Federal Information Processing Standard rating of moderate or high impact.[38] Based on these criteria, we had selected eight mission-essential systems for our review.

Of these systems, the agency had categorized 7 as high-impact systems and 1 as a moderate-impact system. For these 8 selected mission-essential systems, we had reviewed information security program-related controls associated with risk assessments, security plans, security control assessments, remedial action plans, and contingency plans.

To assess the safeguards CDC implemented for its systems, we had examined technical security controls for 24 CDC systems,[39] including systems the agency designated as high-value assets.[40]

---

36 The Center for Internet Security is a nonprofit entity that uses a global information technology community to safeguard private and public organizations against cyber threat. We used the Center for Internet Security tool to assess CDC's information systems.

37 The Interagency Security Committee, an interagency organization chaired by the Department of Homeland Security, was established by Executive Order No. 12977, 60 Fed. Reg. 54411 (October 1995), to enhance the quality and effectiveness of security and the protection of buildings and facilities in the United States occupied by federal employees for nonmilitary activities. Executive Order No. 12977 was later amended by Executive Order No. 13286, 68 Fed. Reg. 10619 (March 2003). The organization is comprised of senior level executives from federal agencies and departments.

38 National Institute of Standards and Technology, *Standards for Security Categorization of Federal Information and Information Systems*, FIPS Publication 199 (Gaithersburg, MD: February 2004). The standard requires agencies to categorize each information system according to the magnitude of harm or impact should the system or its information be compromised. The standard defines three impact levels where the loss of confidentiality, integrity, or availability could be expected to have a limited adverse effect (low), a serious adverse effect (moderate), or a severe or catastrophic adverse effect (high) on organizational operations, organizational assets, or individuals.

39 Because we examined only 24 of the more than 750 systems CDC reported in its FISMA inventory with FIPS 199 categorizations, the results of our review of system-level controls cannot be generalized to the entire CDC environment.

40 High-value assets refer to those assets, systems, facilities, data and datasets that are of particular interest to potential adversaries. These assets, systems, and datasets may contain sensitive controls, instructions or data used in critical federal operations, or house unique collections of data (by size or content), making them of particular interest to criminal,

These included 10 key systems, 8 of which were high- and moderate-impact mission-essential systems just described, 1 additional high-impact system, 1 additional moderate-impact system, and 14 general support systems. We selected the additional high-impact system because the agency re-categorized it as a high-impact system during our review. We selected the additional moderate-impact system because the agency used it to control physical access to highly sensitive CDC biologic lab facilities, including facilities that handle dangerous and exotic substances that cause incurable and deadly diseases.

We selected 10 key systems, 8 of which were mission-essential systems, for review that (1) collect, process, and maintain private or potentially sensitive proprietary business, medical, and personally identifiable information; (2) are essential to CDC's mission; (3) could have a catastrophic or severe impact on operations if compromised; or (4) could be of particular interest to potential adversaries. We also selected 14 general support systems that were part of the agency's network infrastructure supporting the 10 key systems.[41]

To review controls over the 10 key systems and 14 general support systems, we had examined the agency's network infrastructure and assessed the controls associated with system access, encryption, configuration management, and logging and monitoring. For reporting purposes, we had categorized the security controls that we assessed into the five core security functions described in the National Institute of Standards and Technology's (NIST) cybersecurity framework.[42] The five core security functions are:

---

politically-motivated, or state-sponsored actors for either direct exploitation of the data or to cause a loss of confidence in the U.S. government.

41 The 14 general support systems included network devices such as routers, switches, and firewalls; workstations; and servers.

42 National Institute of Standards and Technology, *Framework for Improving Critical Infrastructure Cybersecurity* Version 1.1 (Gaithersburg, MD: Apr. 16, 2018). The framework was developed in response to an executive order issued by the prior administration, *Improving Critical Infrastructure Cybersecurity*, Executive Order 13636 (Washington, D.C.: Feb. 12, 2013). It was originally intended for use in protection of critical infrastructure. NIST initially issued guidance in February 2014 and has since revised the framework.

- Identify: Develop the organizational understanding to manage cybersecurity risk to systems, assets, data, and capabilities.
- Protect: Develop and implement the appropriate safeguards to ensure delivery of critical infrastructure services.
- Detect: Develop and implement the appropriate activities to identify the occurrence of a cybersecurity event.
- Respond: Develop and implement the appropriate activities to take action regarding a detected cybersecurity event.
- Recover: Develop and implement the appropriate activities to maintain plans for resilience and to restore any capabilities or services that were impaired due to a cybersecurity event.

These core security functions are described in more detail in appendix II.

For the *identify* core security function, we had examined CDC's reporting for its hardware and software assets; analyzed risk assessments for the eight selected mission-essential systems to determine whether threats and vulnerabilities were being identified; reviewed risk assessments for two facilities; analyzed CDC policies, procedures, and practices to determine their effectiveness in providing guidance to personnel responsible for securing information and information systems; and analyzed security plans for the eight selected systems to determine if those plans had been documented and updated according to federal guidance. We also evaluated the risk assessments for two facilities that housed the 8 mission-essential selected systems.

For the *protect* core security function, we had examined access controls for the 24 systems. These controls included the complexity and expiration of password settings to determine if password management was being enforced; administrative users' system access permissions to determine whether their authorizations exceeded the access necessary to perform their assigned duties; firewall configurations, among other things, to determine whether system boundaries had been adequately protected; and physical security controls to determine if computer facilities and resources were being protected from espionage, sabotage, damage, and theft.

We also had examined configurations for providing secure data transmissions across the network to determine whether sensitive data were being encrypted. In addition, we had examined configuration settings for routers, network management servers, switches, firewalls, and workstations to determine if settings adhered to configuration standards, and inspected key servers and workstations to determine if critical patches had been installed and/or were up-to-date. Further, we had examined training records to determine if employees and contractors had received security awareness training according to federal requirements, and whether personnel who have significant security responsibilities had received training commensurate with those responsibilities.

For the *detect* core security function, we had analyzed centralized logging and network traffic monitoring capabilities for key assets connected to the network; analyzed CDC's procedures and results for assessing security controls to determine whether controls for the eight selected mission-essential systems had been sufficiently tested at least annually and based on risk. We also had reviewed the agency's implementation of continuous monitoring practices to determine whether the agency had developed and implemented a continuous monitoring strategy to manage its information technology assets and monitor the security configurations and vulnerabilities for those assets.

For the *respond* core security function, we had reviewed CDC's implementation of incident response practices, including an examination of incident tickets for 11 incidents; and had examined the agency's process for correcting identified deficiencies for the eight selected mission-essential systems.

For the *recover* core security function, we had examined contingency plans for eight selected mission-essential systems to determine whether those plans had been developed and tested.

In assessing CDC's controls associated with this function, as well as the other four core functions, we had interviewed Office of the Chief Information Officer officials, as needed.

Within the core security functions, as appropriate, we had evaluated the elements of CDC's information security program based on elements required by FISMA. For example, we analyzed risk assessments, security plans, security control assessments, and remedial action plans for each of the 8 selected mission-essential systems. In addition, we had assessed whether the agency had ensured staff had completed security awareness training and whether those with significant security responsibilities received commensurate training. We also had evaluated CDC's security policies and procedures.

To determine the reliability of CDC's computer-processed data for training and incident response records, we had evaluated the materiality of the data to our audit objective and assessed the data by various means, including reviewing related documents, interviewing knowledgeable agency officials, and reviewing internal controls. Through a combination of methods, we concluded that the data were sufficiently reliable for the purposes of our work.

To accomplish our second objective—on CDC's actions to address the previously identified security program and technical control deficiencies and related recommendations[43]—we requested that the agency provide a status report of its actions to implement each of the recommendations. For each recommendation that CDC indicated it had implemented as of August 3, 2018, we examined supporting documents, observed or tested the associated security control or procedure, and/or interviewed the responsible agency officials to assess the effectiveness of the actions taken to implement the recommendation or otherwise resolve the underlying control deficiency. Based on this assessment and CDC status reports, we defined the status of each recommendation into the following 3 categories:

---

[43] GAO-18-437SU.

- closed-implemented—CDC had implemented the recommendation;
- open-partially implemented—CDC had made progress toward, but had not completed, implementing the recommendation; and
- open-not implemented—CDC had not provided evidence that it had acted to implement the recommendation.

We conducted this performance audit from December 2016 to December 2018 in accordance with generally accepted government auditing standards. Those standards require that we plan and perform the audit to obtain sufficient, appropriate evidence to provide a reasonable basis for our findings and conclusions based on our audit objectives. We believe that the evidence obtained provides a reasonable basis for our findings and conclusions based on our audit objectives.

APPENDIX II: THE NATIONAL INSTITUTE OF STANDARDS AND TECHNOLOGY CYBERSECURITY FRAMEWORK

The National Institute of Standards and Technology's cybersecurity framework consists of five core functions: identify, protect, detect, respond, and recover.[44] Within the five functions are 23 categories and 108 subcategories, as described in the table.

[44] National Institute of Standards and Technology, *Framework for Improving Critical Infrastructure Cybersecurity* Version 1.1 (Gaithersburg, MD: Apr. 16, 2018). The framework was developed in response to an executive order issued by the prior administration, *Improving Critical Infrastructure Cybersecurity*, Executive Order 13636 (Washington, D.C.: Feb. 12, 2013). It was originally intended for use in protection of critical infrastructure. NIST initially issued guidance in February 2014 and has since revised the framework.

Table 4. National Institute of Standards and Technology Framework for Improving Critical Infrastructure Cybersecurity

Category	Subcategory
Identify (ID) core function: Asset Management (ID.AM): The data, personnel, devices, systems, and facilities that enable the organization to achieve business purposes are identified and managed consistent with their relative importance to organizational objectives and the organization's risk strategy.	ID.AM-1: Physical devices and systems within the organization are inventoried.
Identify (ID) core function: Asset Management (ID.AM): The data, personnel, devices, systems, and facilities that enable the organization to achieve business purposes are identified and managed consistent with their relative importance to organizational objectives and the organization's risk strategy.	ID.AM-2: Software platforms and applications within the organization are inventoried.
Identify (ID) core function: Asset Management (ID.AM): The data, personnel, devices, systems, and facilities that enable the organization to achieve business purposes are identified and managed consistent with their relative importance to organizational objectives and the organization's risk strategy.	ID.AM-3: Organizational communication and data flows are mapped.
Identify (ID) core function: Asset Management (ID.AM): The data, personnel, devices, systems, and facilities that enable the organization to achieve business purposes are identified and managed consistent with their relative importance to organizational objectives and the organization's risk strategy.	ID.AM-4: External information systems are catalogued.
Identify (ID) core function: Asset Management (ID.AM): The data, personnel, devices, systems, and facilities that enable the organization to achieve business purposes are identified and managed consistent with their relative importance to organizational objectives and the organization's risk strategy.	ID.AM-5: Resources (e.g., hardware, devices, data, time, personnel, and software) are prioritized based on their classification, criticality, and business value.
Identify (ID) core function: Asset Management (ID.AM): The data, personnel, devices, systems, and facilities that enable the organization to achieve business purposes are identified and managed consistent with their relative importance to organizational objectives and the organization's risk strategy.	ID.AM-6: Cybersecurity roles and responsibilities for the entire workforce and third-party stakeholders (e.g., suppliers, customers, and partners) are established.
Identify (ID) core function: Business Environment (ID.BE): The organization's mission, objectives, stakeholders, and activities are understood and prioritized; this information is used to inform cybersecurity roles, responsibilities, and risk management decisions.	ID.BE-1: The organization's role in the supply chain is identified and communicated.

Category	Subcategory
Identify (ID) core function: Business Environment (ID.BE): The organization's mission, objectives, stakeholders, and activities are understood and prioritized; this information is used to inform cybersecurity roles, responsibilities, and risk management decisions.	ID.BE-2: The organization's place in critical infrastructure and its industry sector is identified and communicated.
Identify (ID) core function: Business Environment (ID.BE): The organization's mission, objectives, stakeholders, and activities are understood and prioritized; this information is used to inform cybersecurity roles, responsibilities, and risk management decisions.	ID.BE-3: Priorities for organizational mission, objectives, and activities are established and communicated.
Identify (ID) core function: Business Environment (ID.BE): The organization's mission, objectives, stakeholders, and activities are understood and prioritized; this information is used to inform cybersecurity roles, responsibilities, and risk management decisions.	ID.BE-4: Dependencies and critical functions for delivery of critical services are established.
Identify (ID) core function: Business Environment (ID.BE): The organization's mission, objectives, stakeholders, and activities are understood and prioritized; this information is used to inform cybersecurity roles, responsibilities, and risk management decisions.	ID.BE-5: Resilience requirements to support delivery of critical services are established for all operating states (e.g., under duress/attack, during recovery, normal operations).
Identify (ID) core function: Governance (ID.GV): The policies, procedures, and processes to manage and monitor the organization's regulatory, legal, risk, environmental, and operational requirements are understood and inform the management of cybersecurity risk.	ID.GV-1: Organizational cybersecurity policy is established and communicated.
Identify (ID) core function: Governance (ID.GV): The policies, procedures, and processes to manage and monitor the organization's regulatory, legal, risk, environmental, and operational requirements are understood and inform the management of cybersecurity risk.	ID.GV-2: Cybersecurity roles and responsibilities are coordinated and aligned with internal roles and external partners.
Identify (ID) core function: Governance (ID.GV): The policies, procedures, and processes to manage and monitor the organization's regulatory, legal, risk, environmental, and operational requirements are understood and inform the management of cybersecurity risk.	ID.GV-3: Legal and regulatory requirements regarding cybersecurity, including privacy and civil liberties obligations, are understood and managed.
Identify (ID) core function: Governance (ID.GV): The policies, procedures, and processes to manage and monitor the organization's regulatory, legal, risk, environmental, and operational requirements are understood and inform the management of cybersecurity risk.	ID.GV-4: Governance and risk management processes address cybersecurity risks.

Table 4. (Continued)

Category	Subcategory
Identify (ID) core function: Risk Assessment (ID.RA): The organization understands the cybersecurity risk to organizational operations (including mission, functions, image, or reputation), organizational assets, and individuals.	ID.RA-1: Asset vulnerabilities are identified and documented.
Identify (ID) core function: Risk Assessment (ID.RA): The organization understands the cybersecurity risk to organizational operations (including mission, functions, image, or reputation), organizational assets, and individuals.	ID.RA-2: Cyber threat intelligence is received from information sharing forums and sources.
Identify (ID) core function: Risk Assessment (ID.RA): The organization understands the cybersecurity risk to organizational operations (including mission, functions, image, or reputation), organizational assets, and individuals.	ID.RA-3: Threats, both internal and external, are identified and documented.
Identify (ID) core function: Risk Assessment (ID.RA): The organization understands the cybersecurity risk to organizational operations (including mission, functions, image, or reputation), organizational assets, and individuals.	ID.RA-4: Potential business impacts and likelihoods are identified.
Identify (ID) core function: Risk Assessment (ID.RA): The organization understands the cybersecurity risk to organizational operations (including mission, functions, image, or reputation), organizational assets, and individuals.	ID.RA-5: Threats, vulnerabilities, likelihoods, and impacts are used to determine risk.
Identify (ID) core function: Risk Assessment (ID.RA): The organization understands the cybersecurity risk to organizational operations (including mission, functions, image, or reputation), organizational assets, and individuals.	ID.RA-6: Risk responses are identified and prioritized.
Identify (ID) core function: Risk Management Strategy (ID.RM): The organization's priorities, constraints, risk tolerances, and assumptions are established and used to support operational risk decisions.	ID.RM-1: Risk management processes are established, managed, and agreed to by organizational stakeholders.
Identify (ID) core function: Risk Management Strategy (ID.RM): The organization's priorities, constraints, risk tolerances, and assumptions are established and used to support operational risk decisions.	ID.RM-2: Organizational risk tolerance is determined and clearly expressed.
Identify (ID) core function: Risk Management Strategy (ID.RM): The organization's priorities, constraints, risk tolerances, and assumptions are established and used to support operational risk decisions.	ID.RM-3: The organization's determination of risk tolerance is informed by its role in critical infrastructure and sector specific risk analysis.

Category	Subcategory
Identify (ID) core function: Supply Chain Risk Management (ID.SC): The organization's priorities, constraints, risk tolerances, and assumptions are established and used to support risk decisions associated with managing supply chain risk. The organization has established and implemented the processes to identify, assess and manage supply chain risks.	ID.SC-1: Cyber supply chain risk management processes are identified, established, assessed, managed, and agreed to by organizational stakeholders.
Identify (ID) core function: Supply Chain Risk Management (ID.SC): The organization's priorities, constraints, risk tolerances, and assumptions are established and used to support risk decisions associated with managing supply chain risk. The organization has established and implemented the processes to identify, assess and manage supply chain risks.	ID.SC-2: Suppliers and third party partners of information systems, components, and services are identified, prioritized, and assessed using a cyber supply chain risk assessment process.
Identify (ID) core function: Supply Chain Risk Management (ID.SC): The organization's priorities, constraints, risk tolerances, and assumptions are established and used to support risk decisions associated with managing supply chain risk. The organization has established and implemented the processes to identify, assess and manage supply chain risks.	ID.SC-3: Contracts with suppliers and third-party partners are used to implement appropriate measures designed to meet the objectives of an organization's cybersecurity program and Cyber Supply Chain Risk Management Plan.
Identify (ID) core function: Supply Chain Risk Management (ID.SC): The organization's priorities, constraints, risk tolerances, and assumptions are established and used to support risk decisions associated with managing supply chain risk. The organization has established and implemented the processes to identify, assess and manage supply chain risks.	ID.SC-4: Suppliers and third-party partners are routinely assessed using audits, test results, or other forms of evaluations to confirm they are meeting their contractual obligations.
Identify (ID) core function: Supply Chain Risk Management (ID.SC): The organization's priorities, constraints, risk tolerances, and assumptions are established and used to support risk decisions associated with managing supply chain risk. The organization has established and implemented the processes to identify, assess and manage supply chain risks.	ID.SC-5: Response and recovery planning and testing are conducted with suppliers and third-party providers.
Protect (PR) core function: Identity Management, Authentication and Access Control (PR.AC): Access to physical and logical assets and associated facilities is limited to authorized users, processes, and devices, and is managed consistent with the assessed risk of unauthorized access to authorized activities and transactions.	PR.AC-1: Identities and credentials are issued, managed, verified, revoked, and audited for authorized devices, users and processes.

Table 4. (Continued)

Category	Subcategory
Protect (PR) core function: Identity Management, Authentication and Access Control (PR.AC): Access to physical and logical assets and associated facilities is limited to authorized users, processes, and devices, and is managed consistent with the assessed risk of unauthorized access to authorized activities and transactions.	PR.AC-2: Physical access to assets is managed and protected.
Protect (PR) core function: Identity Management, Authentication and Access Control (PR.AC): Access to physical and logical assets and associated facilities is limited to authorized users, processes, and devices, and is managed consistent with the assessed risk of unauthorized access to authorized activities and transactions.	PR.AC-3: Remote access is managed.
Protect (PR) core function: Identity Management, Authentication and Access Control (PR.AC): Access to physical and logical assets and associated facilities is limited to authorized users, processes, and devices, and is managed consistent with the assessed risk of unauthorized access to authorized activities and transactions.	PR.AC-4: Access permissions and authorizations are managed, incorporating the principles of least privilege and separation of duties.
Protect (PR) core function: Identity Management, Authentication and Access Control (PR.AC): Access to physical and logical assets and associated facilities is limited to authorized users, processes, and devices, and is managed consistent with the assessed risk of unauthorized access to authorized activities and transactions.	PR.AC-5: Network integrity is protected (e.g., network segregation and network segmentation).
Protect (PR) core function: Identity Management, Authentication and Access Control (PR.AC): Access to physical and logical assets and associated facilities is limited to authorized users, processes, and devices, and is managed consistent with the assessed risk of unauthorized access to authorized activities and transactions.	PR.AC-6: Identities are proofed and bound to credentials and asserted in interactions.
Protect (PR) core function: Identity Management, Authentication and Access Control (PR.AC): Access to physical and logical assets and associated facilities is limited to authorized users, processes, and devices, and is managed consistent with the assessed risk of unauthorized access to authorized activities and transactions.	PR.AC-7: Users, devices, and other assets are authenticated (e.g., single-factor and multi-factor) commensurate with the risk of the transaction (e.g., individuals' security and privacy risks and other organizational risks).
Protect (PR) core function: Awareness and Training (PR.AT): The organization's personnel and partners are provided cybersecurity awareness education and are trained to perform their cybersecurity duties and responsibilities consistent with related policies, procedures, and agreements.	PR.AT-1: All users are informed and trained.

Category	Subcategory
Protect (PR) core function: Awareness and Training (PR.AT): The organization's personnel and partners are provided cybersecurity awareness education and are trained to perform their cybersecurity duties and responsibilities consistent with related policies, procedures, and agreements.	PR.AT-2: Privileged users understand their roles and responsibilities.
Protect (PR) core function: Awareness and Training (PR.AT): The organization's personnel and partners are provided cybersecurity awareness education and are trained to perform their cybersecurity duties and responsibilities consistent with related policies, procedures, and agreements.	PR.AT-3: Third-party stakeholders (e.g., suppliers, customers, partners) understand their roles and responsibilities.
Protect (PR) core function: Awareness and Training (PR.AT): The organization's personnel and partners are provided cybersecurity awareness education and are trained to perform their cybersecurity duties and responsibilities consistent with related policies, procedures, and agreements.	PR.AT-4: Senior executives understand their roles and responsibilities.
Protect (PR) core function: Awareness and Training (PR.AT): The organization's personnel and partners are provided cybersecurity awareness education and are trained to perform their cybersecurity duties and responsibilities consistent with related policies, procedures, and agreements.	PR.AT-5: Physical and cybersecurity personnel understand their roles and responsibilities.
Protect (PR) core function: Data Security (PR.DS): Information and records (data) are managed consistent with the organization's risk strategy to protect the confidentiality, integrity, and availability of information.	PR.DS-1: Data-at-rest is protected.
Protect (PR) core function: Data Security (PR.DS): Information and records (data) are managed consistent with the organization's risk strategy to protect the confidentiality, integrity, and availability of information.	PR.DS-2: Data-in-transit is protected.
Protect (PR) core function: Data Security (PR.DS): Information and records (data) are managed consistent with the organization's risk strategy to protect the confidentiality, integrity, and availability of information.	PR.DS-3: Assets are formally managed throughout removal, transfers, and disposition.
Protect (PR) core function: Data Security (PR.DS): Information and records (data) are managed consistent with the organization's risk strategy to protect the confidentiality, integrity, and availability of information.	PR.DS-4: Adequate capacity to ensure availability is maintained.
Protect (PR) core function: Data Security (PR.DS): Information and records (data) are managed consistent with the organization's risk strategy to protect the confidentiality, integrity, and availability of information.	PR.DS-5: Protections against data leaks are implemented.
Protect (PR) core function: Data Security (PR.DS): Information and records (data) are managed consistent with the organization's risk strategy to protect the confidentiality, integrity, and availability of information.	PR.DS-6: Integrity checking mechanisms are used to verify software, firmware, and information integrity.

Table 4. (Continued)

Category	Subcategory
Protect (PR) core function: Data Security (PR.DS): Information and records (data) are managed consistent with the organization's risk strategy to protect the confidentiality, integrity, and availability of information.	PR.DS-7: The development and testing environment(s) are separate from the production environment.
Protect (PR) core function: Data Security (PR.DS): Information and records (data) are managed consistent with the organization's risk strategy to protect the confidentiality, integrity, and availability of information.	PR.DS-8: Integrity checking mechanisms are used to verify hardware integrity.
Protect (PR) core function: Information Protection Processes and Procedures (PR.IP): Security policies (that address purpose, scope, roles, responsibilities, management commitment, and coordination among organizational entities), processes, and procedures are maintained and used to manage protection of information systems and assets.	PR.IP-1: A baseline configuration of information technology/industrial control systems is created and maintained incorporating security principles (e.g., concept of least functionality).
Protect (PR) core function: Information Protection Processes and Procedures (PR.IP): Security policies (that address purpose, scope, roles, responsibilities, management commitment, and coordination among organizational entities), processes, and procedures are maintained and used to manage protection of information systems and assets.	PR.IP-2: A System Development Life Cycle to manage systems is implemented.
Protect (PR) core function: Information Protection Processes and Procedures (PR.IP): Security policies (that address purpose, scope, roles, responsibilities, management commitment, and coordination among organizational entities), processes, and procedures are maintained and used to manage protection of information systems and assets.	PR.IP-3: Configuration change control processes are in place.
Protect (PR) core function: Information Protection Processes and Procedures (PR.IP): Security policies (that address purpose, scope, roles, responsibilities, management commitment, and coordination among organizational entities), processes, and procedures are maintained and used to manage protection of information systems and assets.	PR.IP-4: Backups of information are conducted, maintained, and tested.
Protect (PR) core function: Information Protection Processes and Procedures (PR.IP): Security policies (that address purpose, scope, roles, responsibilities, management commitment, and coordination among organizational entities), processes, and procedures are maintained and used to manage protection of information systems and assets.	PR.IP-5: Policy and regulations regarding the physical operating environment for organizational assets are met.

Category	Subcategory
Protect (PR) core function: Information Protection Processes and Procedures (PR.IP): Security policies (that address purpose, scope, roles, responsibilities, management commitment, and coordination among organizational entities), processes, and procedures are maintained and used to manage protection of information systems and assets.	PR.IP-6: Data are destroyed according to policy.
Protect (PR) core function: Information Protection Processes and Procedures (PR.IP): Security policies (that address purpose, scope, roles, responsibilities, management commitment, and coordination among organizational entities), processes, and procedures are maintained and used to manage protection of information systems and assets.	PR.IP-7: Protection processes are improved
Protect (PR) core function: Information Protection Processes and Procedures (PR.IP): Security policies (that address purpose, scope, roles, responsibilities, management commitment, and coordination among organizational entities), processes, and procedures are maintained and used to manage protection of information systems and assets.	PR.IP-8: Effectiveness of protection technologies is shared.
Protect (PR) core function: Information Protection Processes and Procedures (PR.IP): Security policies (that address purpose, scope, roles, responsibilities, management commitment, and coordination among organizational entities), processes, and procedures are maintained and used to manage protection of information systems and assets.	PR.IP-9: Response plans (Incident Response and Business Continuity) and recovery plans (Incident Recovery and Disaster Recovery) are in place and managed.
Protect (PR) core function: Information Protection Processes and Procedures (PR.IP): Security policies (that address purpose, scope, roles, responsibilities, management commitment, and coordination among organizational entities), processes, and procedures are maintained and used to manage protection of information systems and assets.	PR.IP-10: Response and recovery plans are tested.
Category	**Subcategory**
Protect (PR) core function: Information Protection Processes and Procedures (PR.IP): Security policies (that address purpose, scope, roles, responsibilities, management commitment, and coordination among organizational entities), processes, and procedures are maintained and used to manage protection of information systems and assets.	PR.IP-11: Cybersecurity is included in human resources practices (e.g., deprovisioning and personnel screening).
Protect (PR) core function: Information Protection Processes and Procedures (PR.IP): Security policies (that address purpose, scope, roles, responsibilities, management commitment, and coordination among organizational entities), processes, and procedures are maintained and used to manage protection of information systems and assets.	PR.IP-12: A vulnerability management plan is developed and implemented **Protect (PR) core function:**.

Table 4. (Continued)

Category	Subcategory
Protect (PR) core function: Protect (PR) core function: Maintenance (PR.MA): Maintenance and repairs of industrial control and information system components are performed consistent with policies and procedures.	PR.MA-1: Maintenance and repair of organizational assets are performed and logged, with approved and controlled tools.
Protect (PR) core function: Maintenance (PR.MA): Maintenance and repairs of industrial control and information system components are performed consistent with policies and procedures.	PR.MA-2: Remote maintenance of organizational assets is approved, logged, and performed in a manner that prevents unauthorized access.
Prot Protect (PR) core function: ective Technology (PR.PT): Technical security solutions are managed to ensure the security and resilience of systems and assets, consistent with related policies, procedures, and agreements.	PR.PT-1: Audit/log records are determined, documented, implemented, and reviewed in accordance with policy.
Protect (PR) core function: Protective Technology (PR.PT): Technical security solutions are managed to ensure the security and resilience of systems and assets, consistent with related policies, procedures, and agreements.	PR.PT-2: Removable media is protected and its use restricted according to policy.
Protect (PR) core function: Protective Technology (PR.PT): Technical security solutions are managed to ensure the security and resilience of systems and assets, consistent with related policies, procedures, and agreements.	PR.PT-3: The principle of least functionality is incorporated by configuring systems to provide only essential capabilities.
Protect (PR) core function: Protective Technology (PR.PT): Technical security solutions are managed to ensure the security and resilience of systems and assets, consistent with related policies, procedures, and agreements.	PR.PT-4: Communications and control networks are protected.
Protect (PR) core function: Protective Technology (PR.PT): Technical security solutions are managed to ensure the security and resilience of systems and assets, consistent with related policies, procedures, and agreements.	PR.PT-5: Mechanisms (e.g., failsafe, load balancing, hot swap) are implemented to achieve resilience requirements in normal and adverse situations.
Detect (DE) core function: Anomalies and Events (DE.AE): Anomalous activity is detected and the potential impact of events is understood.	DE.AE-1: A baseline of network operations and expected data flows for users and systems is established and managed.
Detect (DE) core function: Anomalies and Events (DE.AE): Anomalous activity is detected and the potential impact of events is understood.	DE.AE-2: Detected events are analyzed to understand attack targets and methods.
Detect (DE) core function: Anomalies and Events (DE.AE): Anomalous activity is detected and the potential impact of events is understood.	DE.AE-3: Event data are collected and correlated from multiple sources and sensors.

Category	Subcategory
Detect (DE) core function: Anomalies and Events (DE.AE): Anomalous activity is detected and the potential impact of events is understood.	DE.AE-4: Impact of events is determined.
Detect (DE) core function: Anomalies and Events (DE.AE): Anomalous activity is detected and the potential impact of events is understood.	DE.AE-5: Incident alert thresholds are established.
Detect (DE) core function: Security Continuous Monitoring (DE.CM): The information system and assets are monitored to identify cybersecurity events and verify the effectiveness of protective measures.	DE.CM-1: The network is monitored to detect potential cybersecurity events.
Detect (DE) core function: Security Continuous Monitoring (DE.CM): The information system and assets are monitored to identify cybersecurity events and verify the effectiveness of protective measures	DE.CM-2: The physical environment is monitored to detect potential cybersecurity events.
Detect (DE) core function: Security Continuous Monitoring (DE.CM): The information system and assets are monitored to identify cybersecurity events and verify the effectiveness of protective measures	DE.CM-3: Personnel activity is monitored to detect potential cybersecurity events.
Detect (DE) core function: Security Continuous Monitoring (DE.CM): The information system and assets are monitored to identify cybersecurity events and verify the effectiveness of protective measures	DE.CM-4: Malicious code is detected.
Detect (DE) core function: Security Continuous Monitoring (DE.CM): The information system and assets are monitored to identify cybersecurity events and verify the effectiveness of protective measures	DE.CM-5: Unauthorized mobile code is detected.
Detect (DE) core function: Security Continuous Monitoring (DE.CM): The information system and assets are monitored to identify cybersecurity events and verify the effectiveness of protective measures	DE.CM-6: External service provider activity is monitored to detect potential cybersecurity events.
Detect (DE) core function: Security Continuous Monitoring (DE.CM): The information system and assets are monitored to identify cybersecurity events and verify the effectiveness of protective measures	DE.CM-7: Monitoring for unauthorized personnel, connections, devices, and software is performed.
Detect (DE) core function: Security Continuous Monitoring (DE.CM): The information system and assets are monitored to identify cybersecurity events and verify the effectiveness of protective measures	DE.CM-8: Vulnerability scans are performed.
Detect (DE) core function: Detection Processes (DE.DP): Detection processes and procedures are maintained and tested to ensure awareness of anomalous events.	DE.DP-1: Roles and responsibilities for detection are well defined to ensure accountability.

Table 4. (Continued)

Category	Subcategory
Detect (DE) core function: Detection Processes (DE.DP): Detection processes and procedures are maintained and tested to ensure awareness of anomalous events.	DE.DP-2: Detection activities comply with all applicable requirements.
Detect (DE) core function: Detection Processes (DE.DP): Detection processes and procedures are maintained and tested to ensure awareness of anomalous events.	DE.DP-3: Detection processes are tested.
Detect (DE) core function: Detection Processes (DE.DP): Detection processes and procedures are maintained and tested to ensure awareness of anomalous events.	DE.DP-4: Event detection information is communicated.
Detect (DE) core function: Detection Processes (DE.DP): Detection processes and procedures are maintained and tested to ensure awareness of anomalous events.	DE.DP-5: Detection processes are continuously improved.
Respond (RS) core function: Response Planning (RS.RP): Response processes and procedures are executed and maintained, to ensure response to detected cybersecurity incidents.	RS.RP-1: Response plan is executed during or after an incident.
Respond (RS) core function: Communications (RS.CO): Response activities are coordinated with internal and external stakeholders (e.g., external support from law enforcement agencies).	RS.CO-1: Personnel know their roles and order of operations when a response isneeded.
Respond (RS) core function: Communications (RS.CO): Response activities are coordinated with internal and external stakeholders (e.g., external support from law enforcement agencies).	RS.CO-2: Incidents are reported consistent with established criteria.
Respond (RS) core function: Communications (RS.CO): Response activities are coordinated with internal and external stakeholders (e.g., external support from law enforcement agencies).	RS.CO-3: Information is shared consistent with response plans.
Respond (RS) core function: Communications (RS.CO): Response activities are coordinated with internal and external stakeholders (e.g., external support from law enforcement agencies).	RS.CO-4: Coordination with stakeholders occurs consistent with response plans.
Respond (RS) core function: Communications (RS.CO): Response activities are coordinated with internal and external stakeholders (e.g., external support from law enforcement agencies).	RS.CO-5: Voluntary information sharing occurs with external stakeholders to achieve broader cybersecurity situational awareness.
Respond (RS) core function: Analysis (RS.AN): Analysis is conducted to ensure effective response and support recovery activities.	RS.AN-1: Notifications from detection systems are investigated.

Category	Subcategory
Respond (RS) core function: Analysis (RS.AN): Analysis is conducted to ensure effective response and support recovery activities.	RS.AN-2: The impact of the incident is understood.
Respond (RS) core function: Analysis (RS.AN): Analysis is conducted to ensure effective response and support recovery activities.	RS.AN-3: Forensics are performed.
Respond (RS) core function: Analysis (RS.AN): Analysis is conducted to ensure effective response and support recovery activities.	RS.AN-4: Incidents are categorized consistent with response plans.
Respond (RS) core function: Analysis (RS.AN): Analysis is conducted to ensure effective response and support recovery activities.	RS-AN-5: Processes are established to receive, analyze and respond to vulnerabilities disclosed to the organization from internal and external sources (e.g. internal testing, security bulletins, or security researchers).
Respond (RS) core function: Mitigation (RS.MI): Activities are performed to prevent expansion of an event, mitigate its effects, and eradicate the incident.	RS.MI-1: Incidents are contained.
Respond (RS) core function: Mitigation (RS.MI): Activities are performed to prevent expansion of an event, mitigate its effects, and eradicate the incident.	RS.MI-2: Incidents are mitigated.
Respond (RS) core function: Mitigation (RS.MI): Activities are performed to prevent expansion of an event, mitigate its effects, and eradicate the incident.	RS.MI-3: Newly identified vulnerabilities are mitigated or documented as accepted risks.
Respond (RS) core function: Improvements (RS.IM): Organizational response activities are improved by incorporating lessons learned from current and previous detection/response activities.	RS.IM-1: Response plans incorporate lessons learned.
Respond (RS) core function: Improvements (RS.IM): Organizational response activities are improved by incorporating lessons learned from current and previous detection/response activities.	RS.IM-2: Response strategies are updated.
Recover (RC) core function: Recovery Planning (RC.RP): Recovery processes and procedures are executed and maintained to ensure restoration of systems or assets affected by cybersecurity events.	RC.RP-1: Recovery plan is executed during or after a cybersecurity incident.
Recover (RC) core function: Improvements (RC.IM): Recovery planning and processes are improved by incorporating lessons learned into future activities.	RC.IM-1: Recovery plans incorporate lessons learned.
Recover (RC) core function: Improvements (RS.IM): Organizational response activities are improved by incorporating lessons learned from current and previous detection/response activities.	RC.IM-2: Recovery strategies are updated.

Table 4. (Continued)

Category	Subcategory
Recover (RC) core function: Communications (RC.CO): Restoration activities are coordinated with internal and external parties (e.g., coordinating centers, Internet Service Providers, owners of attacking systems, victims, other CSIRTs, and vendors).	RC.CO-1: Public relations are managed.
Recover (RC) core function: Communications (RC.CO): Restoration activities are coordinated with internal and external parties (e.g., coordinating centers, Internet Service Providers, owners of attacking systems, victims, other CSIRTs, and vendors).	RC.CO-2: Reputation is repaired after an incident.
Recover (RC) core function: Communications (RC.CO): Restoration activities are coordinated with internal and external parties (e.g., coordinating centers, Internet Service Providers, owners of attacking systems, victims, other CSIRTs, and vendors).	RC.CO-3: Recovery activities are communicated to internal and external stakeholders as well as executive and management teams.

Source: National Institute of Standards and Technology. | GAO-19-70.

APPENDIX III: COMMENTS FROM DEPARTMENT OF HEALTH AND HUMAN SERVICES

DEPARTMENT OF HEALTH & HUMAN SERVICES

OFFICE OF THE SECRETARY

Assistant Secretary for Legislation
Washington, DC 20201

DEC 0 6 2018

Gregory C. Wilshusen
Director, Information Security Issues
U.S. Government Accountability Office
441 G Street NW
Washington, DC 20548

Dear Mr. Wilshusen:

Attached are comments on the U.S. Government Accountability Office's (GAO) report entitled, "*Information Security: Significant Progress Made, but CDC Needs to Take Further Action to Resolve Control Deficiencies and Improve Its Program*" (GAO-19-70).

The Department appreciates the opportunity to review this report prior to publication.

Sincerely,

Matthew D. Bassett
Assistant Secretary for Legislation

Attachment

<u>GENERAL COMMENTS FROM THE DEPARTMENT OF HEALTH & HUMAN SERVICES ON THE GOVERNMENT ACCOUNTABILITY OFFICE'S DRAFT REPORT ENTITLED - INFORMATION SECURITY: SIGNIFICANT PROGRESS MADE, BUT CDC NEEDS TO TAKE FURTHER ACTION TO RESOLVE CONTROL DEFICIENCIES AND IMPROVE ITS PROGRAM (GAO-19-70)</u>

CDC greatly appreciates the opportunity to review and comment on this report. CDC recognizes the risks associated with operating a large, global information technology (IT) enterprise and has implemented processes, procedures, and tools to better ensure the prevention, detection, and correction of potential incidents.

Cybersecurity remains a high priority at CDC, and the agency continues to take the responsibility to protect the public health information and data entrusted to it seriously. With a history of operating a robust cybersecurity program, CDC continues to provide sophisticated preventive, detective, and responsive controls to reduce risk to programs, systems, and data. As cyber-threats continue to evolve, the agency recognizes the need for, and remains committed to, protecting public health information and strengthening our cybersecurity posture. CDC has not experienced a major cybersecurity breach that exposed public health information.

CDC appreciates the collaborative interaction with the GAO audit team. Since receiving GAO's report in June 2018, 68 percent of GAO's recommendations have been implemented and work is underway on the remaining 32 percent. As evidenced by these figures, CDC is aggressively working to address all identified deficiencies as quickly and completely as possible.

To further strengthen the agency's cybersecurity program, CDC is restructuring and streamlining the Office of the Chief Information Officer, which includes both the cyber program and IT infrastructure. These changes will increase the agility and resilience of the agency's cybersecurity program. Furthermore, CDC has completed the implementation of phase one and two of the Department of Homeland Security's (DHS) mandatory three-phrase Continuous Diagnostics and Mitigation (CDM) program. CDC leveraged GAO's report to accelerate its implementation, infrastructure, and software deployments for CDM. CDC is actively engaging in phase three, now referred to as CDM DEFEND, and closely following the timeline established by DHS and HHS for completion.

CDC has undertaken unprecedented action through increased investment, dedicated resources, and project management rigor to ensure the protection of CDC information. As a result, tremendous progress has been made on a significant proportion of GAO's recommendations. According to the GAO team, CDC's aggressive and successful approach to recommendation implementation has not been seen in recent audits.

Highlighted below are examples of CDC's efforts and accomplishments:

- Categorized all 184 GAO technical recommendations and 11 program recommendations based on system type, impact level, and level of effort and identified those associated with CDC mission-critical systems to prioritize focus and rapidly reduce risk;
- Used an industry standard and recognized approach to determining risk and applied National Institute of Standards and Technology's (NIST) risk ratings and methodology to all 195 recommendations. This included threat sources maintained by CDC on a routine

GENERAL COMMENTS FROM THE DEPARTMENT OF HEALTH & HUMAN SERVICES ON THE GOVERNMENT ACCOUNTABILITY OFFICE'S DRAFT REPORT ENTITLED - INFORMATION SECURITY: SIGNIFICANT PROGRESS MADE, BUT CDC NEEDS TO TAKE FURTHER ACTION TO RESOLVE CONTROL DEFICIENCIES AND IMPROVE ITS PROGRAM (GAO-19-70)

basis, the likelihood of an adverse event based on previous ability to thwart attacks and current cyber capabilities, and the potential adverse impact to CDC's mission;

- Actively engaged the House Energy and Commerce Committee, initially reporting progress on a bi-weekly basis. Based on significant progress by CDC, this was reduced to monthly at the request of the Committee.
- Provided implementation evidence to GAO for 105 (54 percent) of the technical recommendations by August 8, 2018, of which GAO validated 102 during the initial review of the submissions;
- Provided implementation evidence to GAO for 28 (68 percent) of the high-risk technical recommendations by September 30, 2018;
- Provided implementation evidence to GAO for one of the 11 program recommendations, as of November 1, 2018, and completed significant progress on the remaining ten (10) program recommendations; and
- Executing plans for the remaining, more complex, high-risk recommendations with minimal disruption to CDC's operations, while submitting evidence for an additional 23 recommendations as of November 26, 2018, moving the overall completion rate to 68 percent.

In support of these efforts and accomplishments, CDC:

- Established a Cybersecurity Task Force to strengthen and protect public health computing capabilities to meet program needs;
- Coordinated with all points of contact to verify ownership and track remediation progress;
- Coordinated with key Information Systems Security Officers to verify accountability and responsibility for groups if GAO provided recommendations common to systems, infrastructure, or security domains;
- Developed a plan of action and milestones for all remaining actions not validated by GAO, as of July 23, 2018, to monitor progress for remediation tracking and reporting purposes;
- Provided recurring status updates to CDC leadership in order to communicate remediation progress;
- Created real-time dashboards to track and monitor remediation status in terms of risk, NIST cybersecurity framework categories, and organizational ownership;
- Made critical and appropriate investments to support implementation of necessary IT infrastructure changes and security improvements; and
- Acquired industry-leading expertise to provide support and advisory internal control services necessary to develop actions plans to immediately address GAO audit findings and recommendations. Support also includes execution and program/project management activities to implement the 195 recommendations for improving the cybersecurity program.

GENERAL COMMENTS FROM THE DEPARTMENT OF HEALTH & HUMAN SERVICES ON THE GOVERNMENT ACCOUNTABILITY OFFICE'S DRAFT REPORT ENTITLED - INFORMATION SECURITY: SIGNIFICANT PROGRESS MADE, BUT CDC NEEDS TO TAKE FURTHER ACTION TO RESOLVE CONTROL DEFICIENCIES AND IMPROVE ITS PROGRAM (GAO-19-70)

These efforts, along with other ongoing remediation activities, reinforce CDC's commitment to continuously improve its cybersecurity program to protect its information technology resources and data. Addressing the weaknesses and recommendations identified will strengthen the agency's ability to conduct highly effective incident response, insider threat detection, operational situational awareness, and decrease the overall security risks to sensitive information and IT infrastructure.

CDC is in the midst of remediation activities for all the Program recommendations, and progress to date includes:

Recommendation 1. The Director of CDC should take steps to evaluate system impact level categorizations to ensure that they reflect the current operating environment, particularly for the selected mission essential systems currently categorized as high impact that are interfacing with systems currently categorized at lower impact levels.

CDC concurs with this recommendation and is strengthening processes to ensure that risk assessments for the systems reviewed by GAO reflect the current operating environment, as well as the likelihood and impact of threats to the agency.

Recommendation 2. The Director of CDC should take steps to update risk assessments for selected systems to identify threats, including the likelihood and impact of threats, as well as to address technical constraints and new threats and threat vectors that affect the implementation of controls.

CDC concurs with this recommendation and is strengthening processes to ensure that risk assessments for the systems reviewed by GAO appropriately evaluate the likelihood and impact of threats to the agency, as well as address technical constraints and new threats and threat vectors that affect the implementation of controls.

Recommendation 3. The Director of CDC should take steps to update facility risk assessments for the facilities housing selected systems.

CDC concurs with this recommendation and has updated facility risk assessments for the facilities housing selected systems, which were reviewed by GAO. Evidence for this Program Recommendation is being prepared for submission to GAO for validation.

Recommendation 4. The Director of CDC should take steps to update policies, procedures, and standards with more detailed requirements, to include topics such as (1) configuring and restricting services on network devices and regulating certain connections on firewalls; (2) securely configuring database parameters; and (3) blocking email attachments and deploying a data loss prevention solution.

CDC concurs with this recommendation and is revising the policies, procedures, and standards identified by GAO.

GENERAL COMMENTS FROM THE DEPARTMENT OF HEALTH & HUMAN SERVICES ON THE GOVERNMENT ACCOUNTABILITY OFFICE'S DRAFT REPORT ENTITLED - INFORMATION SECURITY: SIGNIFICANT PROGRESS MADE, BUT CDC NEEDS TO TAKE FURTHER ACTION TO RESOLVE CONTROL DEFICIENCIES AND IMPROVE ITS PROGRAM (GAO-19-70)

Recommendation 5. The Director of CDC should take steps to develop [facility] security plans for the facilities housing resources for the selected systems.

CDC concurs with this recommendation and has developed security plans for the facilities housing selected systems, which were reviewed by GAO, to include all aspects of security for space and equipment, including access control mechanisms, visitor control, and maintenance of records, as well as the process for equipment/inventory control. Evidence for this Program Recommendation is being prepared for submission to GAO for validation.

Recommendation 6. The Director of CDC should take steps to review staff roles and responsibilities to ensure that those with significant security responsibilities are appropriately identified and receive role-based training.

CDC concurs with this recommendation and has implemented a formal process in accordance with National and Departmental standards to monitor and track training requirements for personnel with significant security roles and responsibilities. CDC completed the training requirements by ensuring that all CDC staff with significant security roles and responsibilities have completed role-based training for fiscal year 2018. Evidence for this Program Recommendation has been provided to GAO for validation.

Recommendation 7. The Director of CDC should take steps to routinely monitor the configuration settings of agency systems to ensure that the configurations are set as intended.

CDC concurs with this recommendation and is assessing and enhancing procedures to ensure the monitoring of configuration setting of agency systems are set as intended.

Recommendation 8. The Director of CDC should take steps to update security control assessments for the selected mission essential systems to include an assessment of controls described in system security plans, using an appropriate level of rigor.

CDC concurs with this recommendation and is enhancing processes to increase the rigor with which it performs security assessments on controls of the selected systems reviewed by GAO.

Recommendation 9. The Director of CDC should take steps to update and follow security control assessment plans for the selected mission essential systems, including ensuring the plans have sufficient detail for procedures to be performed and identify assessment team roles and responsibilities.

CDC concurs with this recommendation and is updating and following modified security assessment plans on controls of the selected systems reviewed by GAO, while also ensuring the procedures to be performed and assessment team's roles and responsibilities are clearly defined.

GENERAL COMMENTS FROM THE DEPARTMENT OF HEALTH & HUMAN SERVICES ON THE GOVERNMENT ACCOUNTABILITY OFFICE'S DRAFT REPORT ENTITLED - INFORMATION SECURITY: SIGNIFICANT PROGRESS MADE, BUT CDC NEEDS TO TAKE FURTHER ACTION TO RESOLVE CONTROL DEFICIENCIES AND IMPROVE ITS PROGRAM (GAO-19-70)

Recommendation 10. The Director of CDC should take steps to improve office communication for updating plans of action and milestones for selected systems and act to monitor and support timely completion of corrective actions.

CDC concurs with this recommendation and has partnered with an industry leader to provide support and advisory internal control services necessary to develop actions plans to immediately address GAO audit findings and recommendations. Support also includes execution and program/project management activities to implement the eleven recommendations for improving the cybersecurity program.

Recommendation 11. The Director of CDC should take steps to document the cost-benefit analysis and associated risk of having an alternate processing site within the same geographical region as the main site.

CDC concurs with this recommendation and has developed a cost-benefit analysis, along with all associated risks, of the same geographical region placement of an alternate processing site in relation to the main processing site. Evidence for this recommendation is being prepared for submission to GAO for validation.

CDC appreciates the opportunity to review and comment on our progress associated with this report prior to publication.

APPENDIX V: ACCESSIBLE DATA

Data Tables

Accessible Data for Status of GAO Recommendations to the Centers for Disease Control and Prevention

Open	No Evidence	Open-Partial
72	21	102

Accessible Data for Figure 1: Methods Used by the Centers for Disease Control and Prevention to Assess Security Controls

Interview and examine	Examine	Interview	Test	Unidentified
52	49	38	37	15

Accessible Data for Figure 2: The Centers for Disease Control and Prevention's Planned Timeline for Fully Implementing GAO's Technical Control-Related Recommendations

Month	Number of open recommendations
Mar-2018	184
Apr-2018	184
May-2018	184
Jun-2018	169
Jun-2018	169
Jul-2018	169
Aug-2018	82
Aug-2018	52
Sep-2018	50
Oct-2018	50
Nov-2018	50
Dec-2018	46
Jan-2019	45
Feb-2019	45
Mar-2019	23
Apr-2019	23
May-2019	23
Jun-2019	14
Jul-2019	14
Jul-2019	14
Aug-2019	11
Sep-2019	5
Oct-2019	5
Nov-2019	5
Dec-2019	5
Jan-2020	5
Feb-2020	5
Mar-2020	2
Apr-2020	2
May-2020	2
Jun-2020	2
Jul-2020	2
Aug-2020	2
Sep-2020	2
Sep-2020	2

In: Information Security
Editor: Paul Scholz

ISBN: 978-1-53615-880-9

Chapter 6

INFORMATION SECURITY: AGENCIES NEED TO IMPROVE IMPLEMENTATION OF FEDERAL APPROACH TO SECURING SYSTEMS AND PROTECTING AGAINST INTRUSIONS*

United States Government Accountability Office

ABBREVIATIONS

BOD	binding operational directive
CAP	cross-agency priority
CDM	continuous diagnostics and mitigation
CFO	chief financial officer
CIO	chief information officer

* This is an edited, reformatted and augmented version of United States Government Accountability Office; Report to Congressional Committees, Publication No. GAO-19-105, dated December 2018.

CSIP	*Cybersecurity Strategy and Implementation Plan*
DEFEND	Dynamic and Evolving Federal Enterprise Network Defense
DHS	Department of Homeland Security
DNS	domain name system
FISMA	*Federal Information Security Modernization Act*
IG	inspectors general
IPv6	Internet protocol version 6
NCPS	National Cybersecurity Protection System
NSD	network security deployment
NIST	National Institute of Standards and Technology
OMB	Office of Management and Budget
SCADA	supervisory control and data acquisition
SIEM	security information and event management
SP	special publication
TIC	Trusted Internet Connection
US-CERT	United States Computer Emergency Readiness Team

WHY GAO DID THIS STUDY

Federal agencies are dependent on information systems to carry out operations. The risks to these systems are increasing as security threats evolve and become more sophisticated. To reduce the risk of a successful cyberattack, agencies can deploy intrusion detection and prevention capabilities on their networks and systems.

GAO first designated federal information security as a government-wide high-risk area in 1997. In 2015, GAO expanded this area to include protecting the privacy of personally identifiable information. Most recently, in September 2018, GAO updated the area to identify 10 critical actions that the federal government and other entities need to take to address major cybersecurity challenges.

The federal approach and strategy for securing information systems is grounded in the provisions of the *Federal Information Security*

Modernization Act of 2014 and Executive Order 13800. The act requires agencies to develop, document, and implement an agency-wide program to secure their information systems. The Executive Order, issued in May 2017, directs agencies to use the National Institute of Standards and Technology's cybersecurity framework to manage cybersecurity risks.

The *Federal Cybersecurity Enhancement Act of 2015* contained a provision for GAO to report on the effectiveness of the government's approach and strategy for securing its systems. GAO determined (1) the reported effectiveness of agencies' implementation of the government's approach and strategy; (2) the extent to which DHS and OMB have taken steps to facilitate the use of intrusion detection and prevention capabilities to secure federal systems; and (3) the extent to which agencies reported implementing capabilities to detect and prevent intrusions.

To address these objectives, GAO analyzed OMB reports related to agencies' information security practices including OMB's annual report to Congress for fiscal year 2017. GAO also analyzed and summarized agency-reported security performance metrics and IG-reported information for the 23 civilian CFO Act agencies. In addition, GAO evaluated plans, reports, and other documents related to DHS intrusion detection and prevention programs, and interviewed OMB, DHS, and agency officials.

WHAT GAO RECOMMENDS

GAO is making two recommendations to DHS, to among other things, coordinate with agencies to identify additional needs for training and guidance. GAO is also making seven recommendations to OMB to, among other things, direct the Federal CIO to update the mandated report with required information, such as detecting advanced persistent threats. DHS concurred with GAO's recommendations. OMB did not indicate whether it concurred with the recommendations or not.

What GAO Found

The 23 civilian agencies covered by the *Chief Financial Officers Act of 1990* (CFO Act) have often not effectively implemented the federal government's approach and strategy for securing information systems (see figure below). Until agencies more effectively implement the government's approach and strategy, federal systems will remain at risk. To illustrate:

- As required by Office of Management and Budget (OMB), inspectors general (IGs) evaluated the maturity of their agencies' information security programs using performance measures associated with the five core security functions—identify, protect, detect, respond, and recover. The IGs at 17 of the 23 agencies reported that their agencies' programs were not effectively implemented.
- IGs also evaluated information security controls as part of the annual audit of their agencies' financial statements, identifying material weaknesses or significant deficiencies in internal controls for financial reporting at 17 of the 23 civilian CFO Act agencies.
- Chief information officers (CIOs) for 17 of the 23 agencies reported not meeting all elements of the government's cybersecurity cross-agency priority goal. The goal was intended to improve cybersecurity performance through, among other things, maintaining ongoing awareness of information security, vulnerabilities, and threats; and implementing technologies and processes that reduce malware risk.
- Executive Order 13800 directed OMB, in coordination with the Department of Homeland Security (DHS), to assess and report on the sufficiency and appropriateness of federal agencies' processes for managing cybersecurity risks. Using performance measures for each of the five core security functions, OMB determined that 13 of the 23 agencies were managing overall enterprise risks, while the other 10 agencies were at risk. In assessing agency risk by core

security function, OMB identified a few agencies to be at high risk (see figure at the top of next page).

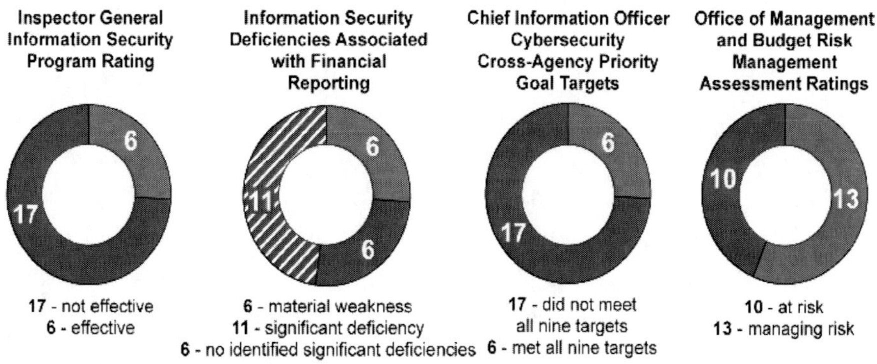

Source: GAO analysis of agency fiscal year 2017 Federal Information Security Modernization Act of 2014 and agency financial reports for fiscal year 2017. | GAO-19-105.

Fiscal Year 2017 Indicators of the 23 Selected Civilian Agencies' Effectiveness in Implementing the Federal Approach and Strategy for Securing Information Systems.

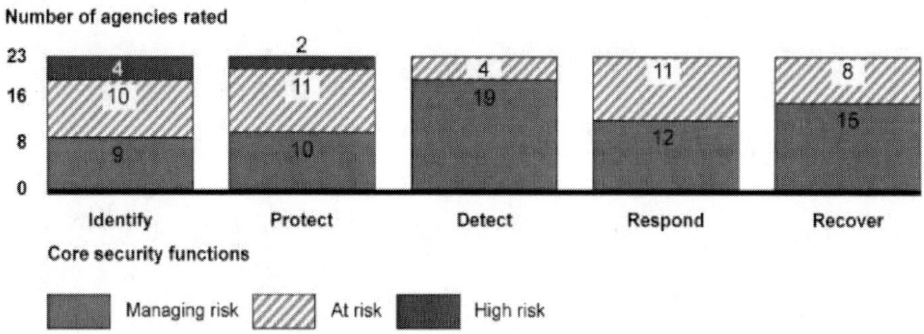

Source: GAO analysis of Office of Management and Budget Fiscal Year 2017 Federal Information Security Modernization Act of 2014 Annual Report to Congress. | GAO-19-105.

Risk Management Assessment Ratings by Core Security Function for the 23 Civilian *Chief Financial Officers Act of 1990* Agencies, Fiscal Year 2017.

DHS and OMB facilitated the use of intrusion detection and prevention capabilities to secure federal agency systems, but further efforts remain. For example, in response to prior GAO recommendations, DHS had improved the capabilities of the National Cybersecurity Protection System (NCPS), which is intended to detect and prevent malicious traffic from entering agencies' computer networks. However, the system still had limitations, such as not having the capability to scan encrypted traffic. The department was also in the process of enhancing the capabilities of federal agencies to automate network monitoring for malicious activity through its Continuous Diagnostics and Mitigation (CDM) program. However, the program was running behind schedule and officials at most agencies indicated the need for additional training and guidance. Further, the Federal CIO issued a mandated report assessing agencies' intrusion detection and prevention capabilities, but the report did not address required information, such as the capability of NCPS to detect advanced persistent threats, and a cost/benefit comparison of capabilities to commercial technologies and tools.

Selected agencies had not consistently implemented capabilities to detect and prevent intrusions into their computer networks. Specifically, the agencies told GAO that they had not fully implemented required actions for protecting email, cloud services, host-based systems, and network traffic from malicious activity. For example, 21 of 23 agencies had not, as of September 2018, sufficiently enhanced email protection through implementation of DHS' directive on enhanced email security. In addition, less than half of the agencies that use cloud services reported monitoring these services. Further, most of the selected 23 agencies had not fully implemented the tools and services available through the first two phases of DHS's CDM program. Until agencies more thoroughly implement capabilities to detect and prevent intrusions, federal systems and the information they process will be vulnerable to malicious threats.

December 18, 2018

The Honorable Ron Johnson
Chairman

The Honorable Claire McCaskill
Ranking Member
Committee on Homeland Security and Governmental Affairs
United States Senate

The Honorable Michael McCaul
Chairman

The Honorable Bennie G. Thompson
Ranking Member
Committee on Homeland Security
House of Representatives

Federal agencies are dependent on computerized (cyber) information systems and electronic data to carry out operations and to process, maintain, and report essential information. Virtually all federal operations are supported by computer systems and electronic data, and agencies would find it difficult, if not impossible, to carry out their missions and account for their resources without these information assets. Hence, the security of these systems and data is vital to public confidence and the nation's safety, prosperity, and well-being. Further, many of these systems contain vast amounts of personally identifiable information,[1] thus, making it imperative to protect the confidentiality, integrity, and availability of this information and effectively respond to data breaches and security incidents when they occur.

The risks to information systems supporting the federal government are increasing as security threats continue to evolve and become more sophisticated. These risks include escalating and emerging threats from around the globe, steady advances in the sophistication of attack technology, and the emergence of new and more destructive attacks.

---

[1] Personally identifiable information is any information that can be used to distinguish or trace an individual's identity, such as name, date and place of birth, or Social Security number, and other types of personal information that can be linked to an individual, such as medical, educational, financial, and employment information.

Compounding these risks, computer networks and systems used by federal agencies are often riddled with security vulnerabilities—both known and unknown. These systems are often interconnected with other internal and external systems and networks, including the Internet, thereby increasing the number of avenues of attack and expanding their attack surface.

Our previous reports, and those by federal inspectors general, describe persistent information security weaknesses that place federal agencies at risk of disruption or inappropriate disclosure of sensitive information. Accordingly, GAO made more than 3,000 recommendations to agencies since 2010 aimed at addressing cybersecurity shortcomings. Although many of these recommendations have been addressed, approximately 700 had not been implemented as of November 2018.

The federal approach and strategy for securing information systems is grounded in the provisions of the *Federal Information Security Modernization Act of 2014* (FISMA)[2] and Executive Order 13800.[3] FISMA requires agencies to develop, document, and implement an agency-wide information security program to secure federal information systems, and assigns oversight responsibilities to the Department of Homeland Security (DHS) and the Office of Management and Budget (OMB). The executive order establishes a policy for managing cybersecurity risk and directs agencies to use the National Institute of Standards and Technology (NIST) cybersecurity framework[4] to manage these risks.

---

[2] The *Federal Information Security Modernization Act of 2014* (FISMA 2014), enacted as Pub. L. No. 113-283, 128 Stat. 3073 (Dec. 18, 2014), largely superseded the *Federal Information Security Management Act of 2002* (FISMA 2002), enacted as *Title III, E-Government Act of 2002*, Pub. L. No. 107-347, 116 Stat. 2899, 2946 (Dec. 17, 2002). As used in this report, FISMA refers both to FISMA 2014 and to those provisions of FISMA 2002 that were either incorporated into FISMA 2014 or were unchanged and continue in full force and effect.

[3] The White House, *Strengthening the Cybersecurity of Federal Networks and Critical Infrastructure*, Executive Order 13800 (Washington, D.C.: May 11, 2017), 82 Fed. Reg. 22391 (May 16, 2017).

[4] National Institute of Standards and Technology, *Framework for Improving Critical Infrastructure Cybersecurity*, Version 1.1 (Gaithersburg, MD: Apr. 16, 2018).

GAO first designated federal information security as a government-wide high-risk area almost 22 years ago in 1997.[5] In 2003,[6] we expanded this area to include computerized systems supporting the nation's critical infrastructure and, in 2015,[7] we further expanded this area to include protecting the privacy of personally identifiable information. We continued to identify federal information security as a government-wide high-risk area in our February 2017 high-risk update report.[8]

Most recently, in September 2018,[9] we provided an update to the information security high-risk area by identifying four major cybersecurity challenges facing the nation and 10 critical actions that the federal government and other entities needed to take to address them. These actions included developing and executing a more comprehensive federal strategy for national cybersecurity and global cyberspace. In this update, we noted that establishing a comprehensive cybersecurity strategy and performing effective oversight were a major challenge.

The *Federal Cybersecurity Enhancement Act of 2015*,[10] which was enacted December 18, 2015, included a provision for GAO to report on the effectiveness of the federal government's approach and strategy for securing agency information systems, including intrusion detection and prevention capabilities.[11] Our specific objectives were to assess: (1) the reported effectiveness of selected agencies' implementation of the federal government's approach and strategy to securing agency information systems; (2) the extent to which OMB and DHS have facilitated the use of intrusion detection and prevention capabilities to secure federal agency

---

[5] GAO, *High-Risk Series: An Overview*, GAO-HR-97-1 (Washington, D.C.: February 1997) and GAO, *High-Risk Series: Information Management and Technology*, GAO-HR-97-9 (Washington, D.C.: February 1997).

[6] GAO, *High Risk Series: An Update*, GAO-03-119 (Washington, D.C.: January 2003).

[7] GAO, *High-Risk Series: An Update*, GAO-15-290 (Washington, D.C.: Feb. 11, 2015).

[8] GAO, *High-Risk Series: Progress on Many High-Risk Areas, While Substantial Efforts Needed on Others*, GAO-17-317 (Washington, D.C.: Feb. 15, 2017).

[9] GAO, *High-Risk Series: Urgent Actions Are Needed to Address Cybersecurity Challenges Facing the Nation*, GAO-18-622, (Washington, D.C.: Sep. 6, 2018).

[10] The act is a part of the *Consolidated Appropriations Act, 2016*, Pub. L. No. 114-113, div. N, title II, subtitle B, 129 Stat. 2242, 2963-2975 (Dec. 18, 2015).

[11] According to NIST, intrusions are defined as attempts to bypass the security mechanisms of a computer or network or to compromise the confidentiality, integrity and availability of the information they contain.

information systems; and (3) the extent to which selected agencies reported implementing capabilities to detect and prevent intrusions.

To address the first objective, we reviewed annual reports from OMB and the 23 civilian agencies covered by the *Chief Financial Officers Act of 1990* (CFO Act).[12] These reports were related to the implementation of FISMA for fiscal year 2017, which was the most recent fiscal year for which the reports were available. In addition, we examined performance metrics related to the cybersecurity cross-agency priority (CAP) goal[13] for fiscal years 2016 and 2017 for the 23 agencies. We also reviewed the financial statement audit reports for the 23 civilian agencies for fiscal years 2016 and 2017. Because we focused our work on the 23 civilian agencies, results from these reviews are not generalizable to the entire federal government.

For the second objective, we collected and reviewed information security-related documents from OMB and DHS and compared them to requirements of the *Federal Cybersecurity Enhancement Act of 2015*. We also interviewed knowledgeable officials from OMB and DHS regarding their agencies' efforts to fulfill requirements of the act.

In addition, we assessed the extent to which DHS had improved the capabilities of the National Cybersecurity Protection System (NCPS).[14] To do this, we assessed the department's actions to implement nine recommendations GAO previously made to, among other things, enhance

---

[12] The 23 civilian *Chief Financial Officers Act of 1990* agencies are the Departments of Agriculture, Commerce, Education, Energy, Health and Human Services, Homeland Security, Housing and Urban Development, the Interior, Justice, Labor, State, Transportation, the Treasury, and Veterans Affairs; the Environmental Protection Agency; General Services Administration; National Aeronautics and Space Administration; National Science Foundation; Nuclear Regulatory Commission; Office of Personnel Management; Small Business Administration; Social Security Administration; and the U.S. Agency for International Development. We did not include the Department of Defense in the scope of our audit because the *Federal Cybersecurity Enhancement Act of 2015* only applies to civilian agencies.

[13] The cybersecurity CAP goal was established by the prior administration as part of implementing the requirement in the *GPRA Modernization Act of 2010*, Pub. L. No. 111-352, § 5, 124 Stat. 3866, 3873 (Jan. 4, 2011) codified at 31 U.S.C. § 1120(a)(1)(B).

[14] NCPS, designed and operated by DHS, was developed to be one of the tools to aid federal agencies in mitigating information security threats. The system is to provide DHS with the capability to provide four cyber-related services to federal agencies: intrusion detection, intrusion prevention, analytics, and information sharing.

the system and better define requirements for future capabilities.[15] We also reviewed documents and interviewed DHS officials to determine other actions, beyond those related to our recommendations, that the department had taken to improve the system. Further, we held semi-structured interviews[16] with knowledgeable officials from the 23 civilian CFO Act agencies to obtain their views on the intrusion detection and prevention capabilities made available by DHS. The results of these interviews are not generalizable to all federal agencies.

To address the third objective, we summarized information from our semi-structured interviews about reported capabilities implemented at the 23 civilian CFO Act agencies to detect and prevent intrusions. We also analyzed security status reports from DHS. See appendix I for additional details on our objectives, scope, and methodology.

We conducted this performance audit from December 2017 to December 2018 in accordance with generally accepted government auditing standards. Those standards require that we plan and perform the audit to obtain sufficient, appropriate evidence to provide a reasonable basis for our findings and conclusions based on our audit objectives. We believe that the evidence obtained provides a reasonable basis for our findings and conclusions based on our audit objectives.

BACKGROUND

Cybersecurity incidents continue to impact federal entities and the information they maintain. According to OMB's 2018 annual FISMA report to Congress, agencies reported 35,277 information security incidents

---

[15] GAO, *Information Security: DHS Needs to Enhance Capabilities, Improve Planning, and Support Greater Adoption of Its National Cybersecurity Protection System,* GAO-16-294 (Washington, D.C.: Jan. 28, 2016).

[16] A semi-structured interview methodology generally involves asking a similar subset of questions of multiple interviewees. We used a semi-structured interview format with both closed- and open-ended questions. The intent of our open-ended questions was to engage the agency officials in a conversation about the topics being discussed.

to DHS's U.S. Computer Emergency Readiness Team (US-CERT)[17] in fiscal year 2017. As shown in Figure 1, these incidents involved threat vectors, such as web-based attacks, phishing attacks,[18] and the loss or theft of computer equipment, among others.[19]

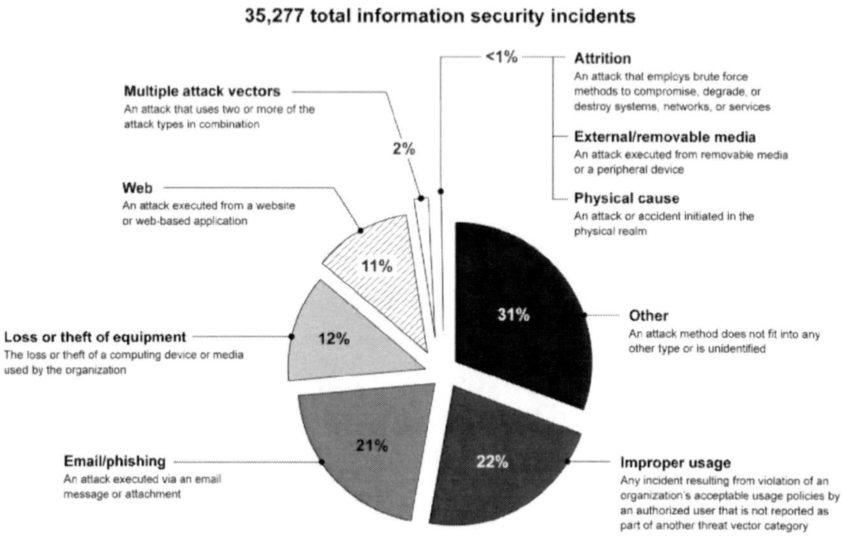

Source: GAO analysis of United States Computer Emergency Readiness Team and Office of Management and Budget data for fiscal year 2017. | GAO-19-105.

Figure 1. Federal Information Security Incidents by Threat Vector Category, Fiscal Year 2017.

These incidents and others like them can pose a serious challenge to economic, national, and personal privacy and security. The following examples highlight the impact of such incidents:

- In March 2018, the Department of Justice reported that it had indicted nine Iranians for conducting a massive cybersecurity theft

[17] Within DHS, US-CERT is a component of the National Cybersecurity and Communications Integration Center. It serves as the central federal information security incident center specified by FISMA.

[18] Phishing is a digital form of social engineering that uses authentic-looking, but fake, emails to request information from users or direct them to a fake website that requests information.

[19] A threat vector (or avenue of attack) specifies the conduit or means used by the source or attacker to initiate a cyberattack.

campaign on behalf of the Islamic Revolutionary Guard Corps. According to the department, the Iranians allegedly stole more than 31 terabytes of documents and data from more than 140 American universities, 30 U.S. companies, and 5 federal government agencies, among other entities.

- In March 2018, a joint alert from DHS and the Federal Bureau of Investigation stated that, since at least March 2016, Russian government actors had targeted U.S. government entities and critical infrastructure sectors, including the energy, nuclear, water, aviation, and critical manufacturing sectors.
- In June 2015, the Office of Personnel Management reported that an intrusion into its systems had affected the personnel records of about 4.2 million current and former federal employees. Then, in July 2015, the agency reported that a separate but related incident had compromised its systems and the files related to background investigations for at least 21.5 million individuals.

Federal Law and Policy Prescribe the Federal Approach and Strategy for Securing Information Systems

The federal approach and strategy for securing information systems is prescribed by federal law and policy. FISMA sets requirements for effectively securing federal systems and information. In addition, the *Federal Cybersecurity Enhancement Act of 2015* requires protecting federal networks through the use of federal intrusion prevention and detection capabilities. Further, Executive Order 13800, *Strengthening the Cybersecurity of Federal Networks and Critical Infrastructure*,[20] directs agencies to manage cybersecurity risks to the federal enterprise by, among other things, using the NIST *Framework for Improving Critical Infrastructure Cybersecurity*[21] (cybersecurity framework).

---

[20] The White House, *Strengthening the Cybersecurity of Federal Networks and Critical Infrastructure*, Executive Order 13800 (Washington, D.C.: May 11, 2017), 82 Fed. Reg. 22391 (May 16, 2017).

[21] National Institute of Standards and Technology, *Framework for Improving Critical Infrastructure Cybersecurity*, Version 1.1 (Gaithersburg, MD: Apr. 16, 2018).

The Federal Information Security Modernization Act of 2014 Sets Requirements for Securing Federal Systems and Information

FISMA was enacted to improve federal cybersecurity and clarify government-wide responsibilities. The law is intended to provide for improved oversight of federal agencies' information security programs. Specifically, the law provides a comprehensive framework for ensuring the effectiveness of information security controls over information resources that support federal operations and assets. The law is also intended to ensure the effective oversight of information security risks, including those throughout civilian, national security, and law enforcement agencies.

FISMA assigns OMB and DHS oversight roles in ensuring federal agencies' compliance with the law. Among other things, FISMA requires OMB to develop and oversee the implementation of policies, principles, standards, and guidelines on information security in federal agencies, except with regard to national security systems. The law also assigns OMB the responsibility of requiring agencies to identify and provide information security protections commensurate with assessments of risk to their information and information systems. The law further requires DHS to administer the implementation of agency information security policies and practices for non-national security information systems, in consultation with OMB, by developing, issuing, and overseeing implementation of binding operational directives;[22] monitoring agency implementation of information security policies and practices; and convening meetings with senior agency officials to help ensure their effective implementation of information security policies and practices, among other things.

FISMA assigned to NIST the responsibility for developing standards and guidelines that include minimum information security requirements. To this end, NIST has issued several publications to provide guidance for agencies in implementing an information security program. For example,

---

[22] Binding operational directives are compulsory directions to agencies in order to safeguard federal information and information systems, are in accordance with OMB guidelines, and may be revised or repealed by the OMB Director. FISMA authorizes DHS to develop and issue binding operational directives to federal agencies and oversee their implementation by agencies. DHS has developed and issued seven binding operational directives, instructing agencies to, among other things, enhance e-mail security by removing certain insecure protocols.

NIST Special Publication (SP) 800-53[23] provides guidance to agencies on the selection and implementation of information security and privacy controls for systems.

FISMA also assigns to the head of each executive branch agency, responsibility for providing information security protections commensurate with the risk and magnitude of harm resulting from unauthorized access, use, disclosure, disruption, modification, or destruction of information systems used or operated by an agency or by a contractor of an agency or other organization on behalf of an agency. The law also delegates to the agency chief information officer (CIO), or comparable official, the authority to ensure compliance with FISMA requirements. The CIO is responsible for designating a senior agency information security officer whose primary duty is information security.

In addition, the law requires agencies to develop, document, and implement an agency-wide information security program to secure federal information systems. Specifically, these information security programs are to provide risk-based protections for the information and information systems that support the operations and assets of the agency. Further, FISMA requires agencies to comply with DHS binding operational directives, OMB policies and procedures, and NIST federal information processing standards.

FISMA also has reporting requirements for OMB and federal agencies. Specifically, OMB is to report annually, in consultation with DHS, on the effectiveness of agency information security policies and practices, including a summary of major agency information security incidents and an assessment of agency compliance with NIST standards. Further, the law requires agencies to report annually to OMB, DHS, certain congressional committees, and the Comptroller General of the United States on the adequacy and effectiveness of their information security policies, procedures, and practices, as well as their compliance with FISMA.

---

[23] National Institute of Standards and Technology, *Security and Privacy Controls for Federal Information Systems and Organizations*, Special Publication 800-53, Revision 4 (Gaithersburg, MD: April 2013).

The Federal Cybersecurity Enhancement Act of 2015 Articulates Requirements for Protecting Federal Networks through the Use of Federal Intrusion Prevention and Detection Capabilities

The *Federal Cybersecurity Enhancement Act of 2015*, among other things, sets forth authority for enhancing federal intrusion prevention and detection capabilities among federal entities. The act contains several provisions for DHS and OMB. Specifically, the act requires that DHS deploy, operate, and maintain capabilities to prevent and detect cybersecurity risks in network traffic traveling to or from an agency's information system. DHS is to make these capabilities available for use by any agency.

In addition, the act requires DHS to improve intrusion detection and prevention capabilities, as appropriate, by regularly deploying new technologies and modifying existing technologies. The act also requires OMB and DHS, in consultation with appropriate agencies, to review and update government-wide policies and programs to ensure appropriate prioritization and use of network security monitoring tools within agency networks, and to brief appropriate congressional committees.

The Executive Order on Strengthening the Cybersecurity of Federal Networks and Critical Infrastructure Directs Agencies to Use the Cybersecurity Framework for Managing Risks

In May 2017, the President signed Executive Order 13800, which sets policy for managing cybersecurity risk as an executive branch enterprise. Specifically, it outlines actions to enhance cybersecurity across federal agencies and critical infrastructure to improve the nation's cyber posture and capabilities against cybersecurity threats. To this end, the order states that the President will hold executive agency heads accountable for managing agency-wide cybersecurity risk and directs each executive agency to use the NIST cybersecurity framework to manage those risks.

The cybersecurity framework, which provides guidance for cybersecurity activities, is based on five core security functions:

- Identify: Develop an organizational understanding to manage cybersecurity risk to systems, people, assets, data, and capabilities.
- Protect: Develop and implement appropriate safeguards to ensure delivery of critical services.
- Detect: Develop and implement appropriate activities to identify the occurrence of a cybersecurity event.[24]
- Respond: Develop and implement appropriate activities to take action regarding a detected cybersecurity incident.
- Recover: Develop and implement appropriate activities to maintain plans for resilience and to restore capabilities or services that were impaired due to a cybersecurity incident.

According to NIST, these five functions should be performed concurrently and continuously to address cybersecurity risk. In addition, when considered together, they provide a high-level, strategic view of the life cycle of an organization's management of cybersecurity risk. Within the five functions are 23 categories and 108 subcategories that include controls for achieving the intent of each function.[25] Appendix II provides a description of the cybersecurity framework categories and subcategories of controls.

GAO Has Reported on Challenges Related to Establishing a Comprehensive Cybersecurity Strategy

In February 2013, we reported that the government had issued a variety of strategy-related documents that addressed priorities for enhancing cybersecurity within the federal government, as well as for

[24] Cybersecurity events are cybersecurity changes that may have an impact on the organizational operations (including mission, capabilities, or reputation).

[25] For example, "risk assessment" is one of five categories that comprise the "identify" function. The risk assessment category is divided into six subcategories that involve activities such as identifying and documenting internal and external threats; identifying potential business impacts and likelihoods; and determining risk based on threats, vulnerabilities, likelihoods, and impacts. Each subcategory activity cross-references information system controls from various information security publications, including NIST Special Publication 800-53.

encouraging improvements in the cybersecurity of critical infrastructure within the private sector. However, we noted that no overarching cybersecurity strategy had been developed that articulated priority actions, assigned responsibilities for performing them, and set time frames for their completion.[26] Accordingly, we recommended that the White House Cybersecurity Coordinator[27] in the Executive Office of the President develop an overarching federal cybersecurity strategy that included all key elements of the desirable characteristics of a national strategy.[28] These characteristics would include, among other things, milestones and performance measures for major activities to address stated priorities; cost and resources needed to accomplish stated priorities; and specific roles and responsibilities of federal organizations related to the strategy's stated priorities.

Since that time, the executive branch has made progress toward outlining a federal strategy for confronting cyber threats. For example, in September 2018,[29] we reported that recent executive branch initiatives that identify cybersecurity priorities for the federal government provide a good foundation toward establishing a more comprehensive strategy. Nevertheless, we pointed out that additional efforts were needed to address all of the desirable characteristics of a national strategy that we recommended. Specifically, recently issued executive branch strategy documents[30] did not include key elements of desirable characteristics that can enhance the usefulness of a national strategy as guidance for decision makers in allocating resources, defining policies, and helping to ensure

---

[26] GAO, *Cybersecurity: National Strategy, Roles, and Responsibilities Need to Be Better Defined and More Effectively Implemented,* GAO-13-187 (Washington, D.C.: Feb. 14, 2013).

[27] In December 2009, a Special Assistant to the President was appointed as Cybersecurity Coordinator to address the recommendations made in the Cyberspace Policy Review, including coordinating interagency cybersecurity policies and strategies and developing a comprehensive national strategy to secure the nation's digital infrastructure.

[28] In 2004, we developed a set of desirable characteristics that can enhance the usefulness of national strategies in allocating resources, defining policies, and helping to ensure accountability. See GAO, *Combating Terrorism: Evaluation of Selected Characteristics in National Strategies Related to Terrorism*, GAO-04-408T (Washington, D.C.: Feb. 3, 2004).

[29] GAO-18-622.

[30] These initiatives include Executive Order 13800, the National Security Strategy, and DHS Cybersecurity Strategy.

accountability. For example, these strategy documents did not generally include:

- milestones and performance measures to gauge results;
- resources needed to carry out the goals and objectives; and
- clearly defined roles and responsibilities for key agencies, such as DHS, the Department of Defense, and OMB.

Ultimately, we determined that a more clearly defined, coordinated, and comprehensive approach to planning and executing an overall strategy would likely lead to significant progress in furthering strategic goals and lessening persistent weaknesses.

Subsequent to our September 2018 report, the President issued the *National Cyber Strategy* on September 20, 2018.[31] The strategy builds upon Executive Order 13800 and describes actions that federal agencies and the administration are to take to, among other things, secure federal information systems. For example, the strategy states that the administration is expected to further enable DHS to secure federal department and agency networks, to include ensuring that DHS has appropriate access to agency information systems for cybersecurity purposes and can take and direct action to safeguard systems. In addition, the strategy states that the administration plans to continue with its existing efforts underway to transition agencies to shared services and infrastructure and that DHS is to have appropriate visibility into those services and infrastructure to improve cybersecurity posture.[32]

DHS Offers Federal Agencies Capabilities Intended to Detect and Prevent Intrusions to Federal Information Systems

DHS's Network Security Deployment (NSD) division manages cybersecurity programs that are intended to improve the cybersecurity

[31] The White House, *National Cyber Strategy* (Washington, D.C.: Sept. 20, 2018).

[32] Evaluating the *National Cyber Strategy* to determine if it included the key elements of desirable characteristics discussed in GAO-18-622 was not within the scope of this review.

posture of the federal government. Among these programs, NCPS provides a capability to detect and prevent potentially malicious network traffic from entering agencies' networks. In addition, the Continuous Diagnostics and Mitigation (CDM) program provides tools to agencies intended to identify and resolve cyber vulnerabilities on an ongoing basis.

DHS's National Cybersecurity Protection System Is Intended to Detect and Prevent Cyber Intrusions

Table 1. Overview of the National Cybersecurity Protection System (NCPS) Deployment, 2003-2013

Operational name	Deployment year	NCPS objective	Description
EINSTEIN 1	2003	Intrusion detection	Provides an automated process for collecting, correlating, and analyzing agencies' computer network traffic information from sensors installed at their Internet connections.[a]
EINSTEIN 2	2009	Intrusion detection	Monitors federal agency Internet connections for specific predefined signatures of known malicious activity and alerts DHS's U.S. Computer Emergency Readiness Team (US-CERT) when specific network activity matching the predetermined signatures is detected.[b]
EINSTEIN 3 Accelerated	2013	Intrusion detection Intrusion prevention	Automatically blocks malicious traffic from entering or leaving federal civilian agency networks. This capability is managed by Internet service providers, who administer intrusion prevention and threat-based decision making using DHS-developed indicators of malicious cyber activity to develop signatures.[c]

Source: GAO analysis of Department of Homeland Security (DHS) data. | GAO-19-105.

[a]The network traffic information includes source and destination Internet Protocol addresses used in the communication, source and destination ports, the time the communication occurred, and the protocol used to communicate.

[b]Signatures are recognizable, distinguishing patterns associated with cyberattacks, such as a binary string associated with a computer virus or a particular set of keystrokes used to gain unauthorized access to a system.

[c]An indicator is defined by DHS as human-readable cyber data used to identify some form of malicious cyber activity. These data may be related to Internet Protocol addresses, domains, e-mail headers, files, and character strings. Indicators can be either classified or unclassified.

Operated by DHS's US-CERT, NCPS is intended to detect and prevent cyber intrusions into agency networks, analyze network data for trends and anomalous data, and share information with agencies on cyber threats and incidents. Deployed in stages, this system, operationally known as EINSTEIN, has provided increasing capabilities to detect and prevent potential cyberattacks involving the network traffic entering or exiting the networks of participating federal agencies. Table 1 provides an overview of the EINSTEIN deployment stages to date.

In January 2016, we reported the projected total life-cycle cost of the program was approximately $5.7 billion through fiscal year 2018.[33] In addition, according to the Federal CIO, Congress appropriated $468 million in fiscal year 2017 and $402 million in fiscal year 2018 for NCPS.

In that report, we also noted that NCPS was partially, but not fully, meeting most of its stated system objectives.[34] Although the system's intrusion detection capabilities provided the ability to detect known patterns of malicious activity on agency networks, it was limited in its capabilities to identify potential threats using anomaly-based detection. We also reported that although DHS had developed metrics for measuring the performance of NCPS, the metrics did not gauge the quality, accuracy, or effectiveness of the system's intrusion detection and prevention capabilities.

The department had also identified needs for future capabilities, but had not defined requirements for the capability to detect threats entering and exiting cloud service providers. Further, DHS had not considered specific vulnerability information for agency information systems in making risk-based decisions about future intrusion prevention capabilities.

Accordingly, we made nine recommendations to DHS to, among other things, enhance the NCPS capabilities for meeting its objectives and better define requirements for future capabilities. DHS agreed with each of our nine recommendations and indicated that it would take steps to address them.

---

33 GAO-16-294.

34 GAO-16-294.

DHS's Continuous Diagnostics and Mitigation Program Provides Agencies with Tools and Services Intended to Secure Agency Systems

DHS's CDM program provides federal agencies with tools and services that have the intended capability to automate network monitoring, correlate and analyze security-related information, and enhance risk-based decision making at agency and government-wide levels. These tools include sensors that perform automated scans or searches for known cyber vulnerabilities, the results of which can feed into a dashboard that, at an agency level, is intended to alert network managers and enable the agency to allocate resources based on the risk. Summary data from each participating agency's dashboard is expected to be transmitted to the Federal Dashboard where the data can be used to inform decisions about cybersecurity risks across the federal government.

There are four phases of CDM implementation:

- Phase 1—involves deploying products to automate hardware and software asset management, configuration settings, and common vulnerability management capabilities. According to the *Cybersecurity Strategy and Implementation Plan*,[35] DHS purchased phase 1 tools and integration services for all participating agencies in fiscal year 2015. DHS plans to have all phase 1 tools deployed at participating agencies by the end of the second quarter of fiscal year 2019.
- Phase 2—intends to address privilege management and infrastructure integrity by allowing agencies to monitor users on their networks and to detect whether users are engaging in unauthorized activity. According to the *Cybersecurity Strategy and Implementation Plan*, DHS was to provide agencies with additional phase 2 capabilities throughout fiscal year 2016, with the full suite of CDM phase 2 capabilities delivered by the end of

---

[35] Office of Management and Budget, *Cybersecurity Strategy and Implementation Plan (CSIP) for the Federal Civilian Government*, Memorandum M-16-04 (Washington, D.C.: Oct. 30, 2015). CSIP identified objectives, key actions, responsibilities, and timeframes for completing actions that were intended to strengthen cybersecurity at federal civilian agencies.

that fiscal year. However, according to the OMB FISMA *Annual Report to Congress for Fiscal Year 2017*, the CDM program began deploying Phase 2 tools and sensors during fiscal year 2017.[36] DHS plans to have all phase 2 tools deployed at participating agencies by the end of fiscal year 2019.

- Phase 3—includes detection capabilities that are intended to assess agency network activity and identify any anomalies that may indicate a cybersecurity compromise.[37] Full operating capability[38] for phases 1, 2, and 3 is planned to be achieved by the end of fiscal year 2022.[39]
- Phase 4—intends to provide tools to (1) protect data at rest, in transit, and in use; (2) prevent loss of data; and (3) manage and mitigate data breaches. According to CDM program officials, phase 4 has not been approved and no tools have been selected.

NIST Recommends That Federal Agencies Deploy Intrusion Detection and Prevention Capabilities

An approach for protecting systems against cybersecurity compromise is for federal agencies to build successive layers of defense mechanisms at strategic points in their information technology infrastructures. This approach, commonly referred to as defense in depth, entails implementing a series of protective mechanisms so that if one mechanism fails to detect and prevent an attack, another will provide a backup defense. By utilizing defense in depth, federal agencies can reduce the risk of a successful

---

[36] Office of Management and Budget, *Federal Information Security Modernization Act of 2014, Annual Report to Congress Fiscal Year 2017* (Washington, D.C.: Mar. 2018).

[37] CDM phase 3 also includes, among other things, completing remaining required phase 1 activities and integrating all deployed phase 2 tools into agency and federal dashboards.

[38] DHS's NSD division considers an agency to have reached full operating capability for a given CDM phase when the full set of capabilities for the phase has been fully deployed across the agency (i.e., the capabilities have been installed, configured, integrated, and data is feeding into the agency's dashboard).

[39] According to CDM program officials, initial operational capability—a project milestone attained when the capabilities for a phase have been fully deployed to at least five agencies—for phase 3 is planned by the end of fourth quarter fiscal year 2019.

cyberattack by implementing intrusion detection and prevention capabilities.

NIST has developed guidelines for protecting agency information systems using intrusion detection and prevention capabilities. For example, NIST SP 800-53 recommends that agencies strategically deploy capabilities and perform monitoring of their systems to include observation of events occurring on their network and at the external boundary of their network. In addition, NIST SP 800-94 provides agencies with guidance in designing, implementing, configuring, securing, monitoring, and maintaining such capabilities.[40]

As part of their defense-in-depth approach and, as recommended by the NIST guidelines, agencies can deploy the following list of capabilities, among others, on their networks to detect and prevent an attack:

- Protecting email from intrusions: According to OMB,[41] email, by way of phishing attacks, remains one of the most common threat vectors across the government. Methods for protecting email include encryption, false email alerts, and anti-spear-phishing training.[42]
- Monitoring cloud services: Cloud vendors provide services to agencies, including Software as a Service,[43] Platform as a Service,[44] and Infrastructure as a Service.[45] As agencies

---

[40] National Institute of Standards and Technology, *Guide to Intrusion Detection and Prevention Systems*, Special Publication 800-94 (Gaithersburg, MD: February 2007).

[41] Office of Management and Budget, *Federal Cybersecurity Risk Determination Report and Action Plan* (Washington, D.C.: May 2018).

[42] Spear phishing represents a digital form of social engineering that uses authentic looking emails, websites, or instant messages that are closely tailored to their intended audience to get users to download malware, open malicious attachments, or open links that direct them to a website that requests information or executes malicious code.

[43] In Software as a Service, the agency uses the service provider's applications, which are accessible from various client devices through an interface such as a web browser (e.g., web-based e-mail system). The agency does not manage or control the underlying infrastructure or the individual application capabilities.

[44] In Platform as a Service, the agency deploys its own or acquired applications created using programming languages and tools supported by the provider. The agency does not manage or control the underlying infrastructure, but controls and configures the deployed applications.

increasingly rely on cloud services, monitoring traffic to and from their cloud service providers helps to ensure that agencies detect malicious traffic.

- Using host-based intrusion prevention: Host-based intrusion prevention systems provide defense at an individual system or device level by protecting against malicious activities. Host-based capabilities include memory-based protection[46] and application whitelisting.[47]
- Monitoring external and internal traffic: Agencies can monitor external and internal traffic, including: encrypted traffic, traffic between workstations and servers on the network, and direct connections to outside entities such as universities. Monitoring traffic helps to ensure that agencies detect malicious activity.
- Using security information and event management: A security information and event management capability produces real-time alerts and notifications of significant security events. Security alerts and notifications can provide the agency with better situational awareness regarding possible intrusion activity.

SELECTED AGENCIES WERE NOT EFFECTIVELY IMPLEMENTING THE FEDERAL GOVERNMENT'S APPROACH AND STRATEGY TO SECURING INFORMATION SYSTEMS

According to inspectors general, agency CIOs, and OMB reports on federal information security practices, many agencies were not effectively

---

[45] In Infrastructure as a Service, the agency has the capability to provision processing, storage, networks, and other fundamental computing resources and run its own software, including operating systems and applications. The agency does not manage or control the underlying infrastructure but controls and configures operating systems, storage, deployed applications, and possibly, selected networking components (e.g., host firewalls).

[46] Memory based protections are safeguards that protect memory from unauthorized code execution.

[47] An application whitelist is a list of applications and application components that an agency has authorized for use on its hosts.

implementing the federal government's approach and strategy to securing information systems as of fiscal year 2017. Agencies' inspectors general determined that most of the 23 civilian CFO Act agencies did not have effective agency-wide information security programs. They also reported that agencies did not have effective information security controls in place, leading to deficiencies in internal control over financial reporting. In addition, the CIOs demonstrated that, during fiscal years 2016 and 2017, most agencies had not met all targets for the cybersecurity CAP goal for improving cybersecurity performance.

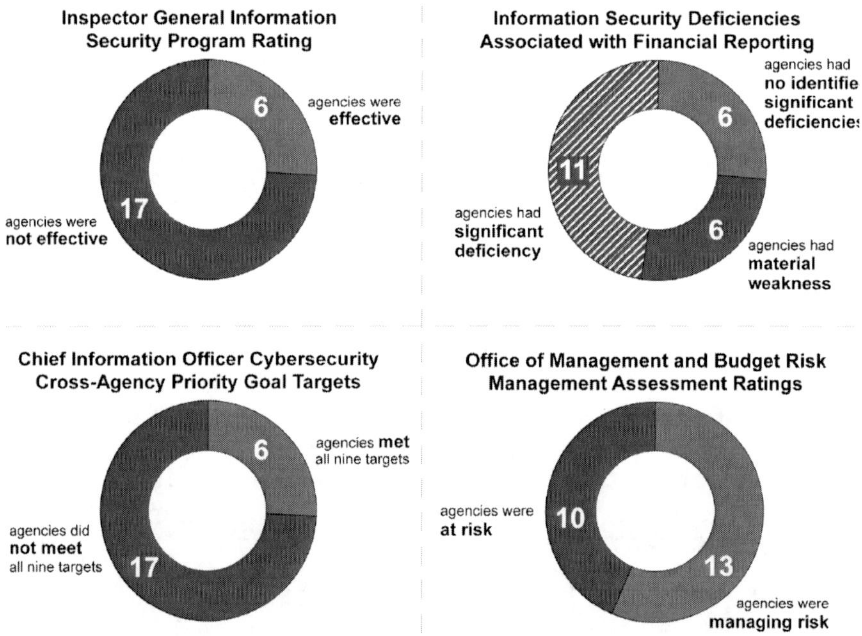

Source: GAO analysis of agency fiscal year 2017 Federal Information Security Modernization Act of 2014 and agency financial reports for fiscal year 2017. | GAO-19-105.

Figure 2. Fiscal Year 2017 Indicators of the 23 Civilian *Chief Financial Officers Act of 1990* Agencies' Effectiveness in Implementing the Federal Approach and Strategy for Securing Information Systems.

Further, based on FISMA metrics reported for fiscal year 2017, OMB determined that 13 of the 23 agencies were managing risks to their enterprise, while the other 10 agencies were at risk of ineffectively identifying, protecting, detecting, responding to, and if necessary, recovering from cyber intrusions. Figure 2 summarizes agencies' efforts to implement the government's approach and strategy for securing information systems as of fiscal year 2017. Appendix III includes a table that provides an additional overview of the effectiveness of each agency's implementation of the government's approach and strategy to securing information systems.

Inspectors General Determined That Most Selected Agencies Did Not Have Effective Information Security Programs or Controls in Place as of Fiscal Year 2017

Inspectors general determined that more than half of the 23 civilian CFO Act agencies did not have effective agency-wide information security programs as of fiscal year 2017. Further, in agency financial statement audit reports for fiscal year 2017, inspectors general reported that, despite improvements being made in information security practices, most of the civilian CFO Act agencies continued to exhibit deficiencies in information security controls. As a result of these deficiencies, inspectors general reported material weaknesses or significant deficiencies in internal control over financial reporting.

Inspectors General Indicate That Few Agencies Had Effective Information Security Programs

FISMA requires inspectors general to determine the effectiveness of their respective agencies' information security programs. To do so, FISMA reporting instructions[48] direct inspectors general to provide a maturity

[48] Inspectors general FISMA metrics and reporting instructions were developed as a collaborative effort amongst OMB, DHS, and the Council of the Inspectors General on Integrity and Efficiency, in consultation with the Federal CIO Council. The FISMA metrics and reporting instructions provide reporting requirements across key areas to be addressed in the independent assessment of agencies information security programs. See *Fiscal Year 2017 Inspector General Federal Information Security Modernization Act of 2014 Reporting Metrics* (April 17, 2017).

rating for agency information security policies, procedures, and practices related to the five core security functions established in the NIST cybersecurity framework, as well as for the agency-wide information security program.

The ratings used to evaluate the effectiveness of agencies' information security programs are based on a five-level maturity model, as described in Table 2.

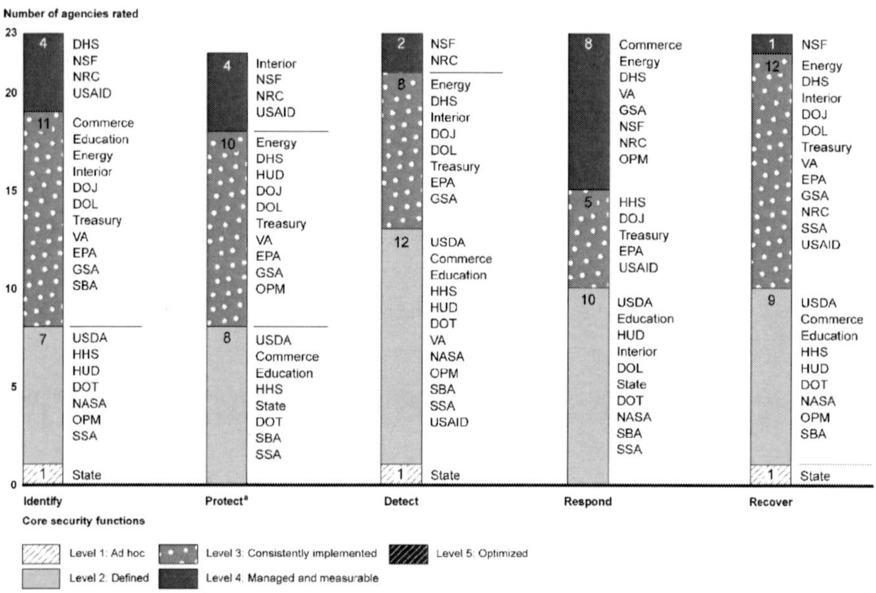

DHS (Department of Homeland Security), NSF (National Science Foundation), NRC (Nuclear Regulatory Commission), USAID (United States Agency for International Development), Commerce (Department of Commerce), Education (Department of Education), Energy (Department of Energy), Interior (Department of the Interior), DOJ (Department of Justice), DOL (Department of Labor), Treasury (Department of the Treasury), VA (Department of Veterans Affairs), EPA (Environmental Protection Agency), GSA (General Services Administration), SBA (Small Business Administration), USDA (United States Department of Agriculture), HHS (Department of Health & Human Services), HUD (Department of Housing and Urban Development), DOT (Department of Transportation), NASA (National Aeronautics and Space Administration), OPM (Office of Personnel Management), SSA (Social Security Administration), State (Department of State)

Source: GAO analysis of agency fiscal year 2017 Federal Information Security Modernization Act of 2014 reports. | GAO-19-105.

[a]Only 22 agencies are listed because the NASA inspector general did not provide a rating for the Protect function.

Figure 3. Inspector General Ratings of Agencies' Information Security Policies, Procedures, and Practices Related to the Five Core Security Functions, as of Fiscal Year 2017.

According to this maturity model, Level 4 (managed and measurable) represents an effective level of security.[49] Therefore, if an inspector general rates the agency's information security program at Level 4 or Level 5, then that agency is considered to have an effective information security program.[50]

For fiscal year 2017, the inspectors general for 6 of the 23 civilian CFO Act agencies reported that their agencies had an effective agency-wide information security program. More specifically, for the 5 core security functions, most inspectors general reported that their agency was at Level 3 (consistently implemented) for the *identify*, *protect*, and *recover* functions, and at Level 2 (defined) for the *detect* and *respond* functions, as shown in Figure 3.

Table 2. Inspector General Reporting Metrics Maturity Model

Maturity level	Description
Level 1: Ad hoc	Policies, procedures, and strategy are not formalized; activities are performed in an ad hoc, reactive manner.
Level 2: Defined	Policies, procedures, and strategy are formalized and documented, but not consistently implemented.
Level 3: Consistently Implemented	Policies, procedures, and strategy are consistently implemented, but quantitative and qualitative effectiveness measures are lacking.
Level 4: Managed and Measurable	Quantitative and qualitative measures on the effectiveness of policies, procedures, and strategy are collected across the organization and used to assess those policies procedures, and strategy, and make necessary changes.
Level 5: Optimized	Policies, procedures, and strategy are fully institutionalized, repeatable, self-generating, consistently implemented, and regularly updated based on a changing threat and technology landscape and business/mission needs.

Source: GAO analysis of Fiscal Year 2017 Inspector General *Federal Information Security Modernization Act of 2014* Reporting Metrics, April 17, 2017. | GAO-19-105.

[49] NIST defines security control effectiveness as the extent to which security controls are implemented correctly, operate as intended, and produce the desired outcome with respect to meeting the security requirements for the information system and are in compliance with established security policies.

[50] *Fiscal Year 2017 Inspector General Federal Information Security Modernization Act of 2014 Reporting Metrics* (April 17, 2017).

Inspectors General Continued to Identify Significant Security Control Deficiencies in Controls over Financial Reporting at Most Selected Agencies

Inspectors general report on the effectiveness of agencies' information security controls as part of the annual audits of the agencies' financial statements. The reports resulting from these audits include a description of information security control deficiencies related to the five major control categories defined by the *Federal Information System Controls Audit Manual* (FISCAM)—access controls, configuration management, segregation of duties, contingency planning, and security management.[51] The reports also identify the inspectors general's designation of information security as a significant deficiency[52] or material weakness in internal control over financial reporting systems.[53]

For fiscal year 2017, inspectors general continued to identify information security control deficiencies in each of the five major control categories across the 23 civilian CFO Act agencies. The number of agencies with deficiencies in the access control and contingency planning

---

[51] FISCAM is GAO's audit methodology for performing information system control audits in accordance with generally acceptable government auditing standards. The five control categories defined by this manual include: (1) access controls that limit or detect access to computer resources, thereby protecting them against unauthorized modification, loss, and disclosure; (2) configuration management controls that prevent unauthorized changes to information system resources and to assure that software is current and known vulnerabilities are patched; (3) segregation of duties controls that prevent an individual from controlling all critical stages of a process by splitting responsibilities between two or more organizational groups; (4) contingency planning controls that help avoid significant disruptions in computer-dependent operations; (5) and security management controls that provide a framework for ensuring that risks are understood and that effective controls are selected, implemented, and operating as intended. See GAO, *Federal Information System Controls Audit Manual* (FISCAM), GAO-09-232G (Washington, D.C.: February 2009).

[52] A significant deficiency is a deficiency, or a combination of deficiencies, in internal control over financial reporting that is less severe than a material weakness, yet important enough to merit attention by those charged with governance. A deficiency in internal control exists when the design or operation of a control does not allow management or employees, in the normal course of performing their assigned functions, to prevent, or detect and correct, misstatements on a timely basis.

[53] A material weakness is a deficiency, or combination of deficiencies, in internal control over financial reporting, such that there is a reasonable possibility that a material misstatement of the entity's financial statements will not be prevented, or detected and corrected, on a timely basis.

information security control categories decreased between fiscal years 2016 and 2017, according to the inspectors general.

Nevertheless, the inspectors general reported that agencies continued to exhibit deficiencies in these two control categories. In addition, the number of agencies with deficiencies in the security management and segregation of duties control categories increased from the prior year. The number of agencies reported as having deficiencies in the configuration management control category remained the same. Figure 4 shows the number of agencies that reported deficiencies in each of the information security control categories for fiscal years 2016 and 2017.

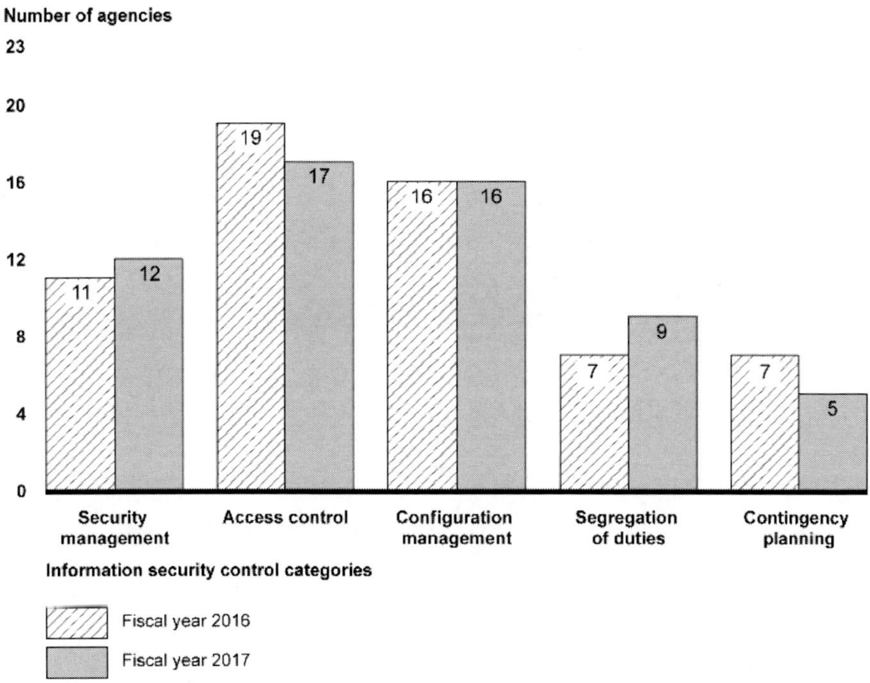

Source: GAO analysis of agency financial reports for fiscal years 2016 and 2017. | GAO-19-105.

Figure 4. Number of Civilian *Chief Financial Officers Act of 1990* Agencies Reporting Deficiencies in Information Security Control Categories for Fiscal Years 2016 and 2017.

Overall, inspectors general for the 23 civilian CFO Act agencies reported progress in agencies' information security practices for fiscal year 2017. Specifically, during that time, 17 inspectors general designated information security as either a significant deficiency (11) or material weakness (6) in internal control over financial reporting systems for their agencies. This is a decrease from the previous fiscal year when 19 inspectors general designated information security as a significant deficiency (12) or material weakness (7).

Most Agencies Reported Not Meeting All Targets for the Cybersecurity Cross-Agency Priority Goal in Fiscal Years 2016 and 2017

Reporting instructions contained in the fiscal year 2017 FISMA metrics[54] directed CIOs to assess their agencies' progress toward achieving outcomes that strengthen federal cybersecurity. To do this, CIOs evaluated their agencies' performance in reaching targets for specific FISMA reporting metrics. According to the reporting instructions, certain metrics were selected to represent the administration's cybersecurity CAP goal. These selected metrics allowed CIOs to evaluate their agencies progress in meeting targets for that goal.

The cybersecurity CAP goal for fiscal years 2015 through 2017[55] was to improve cybersecurity performance by having an ongoing awareness of information security, vulnerabilities, and threats impacting the operating information environment; ensuring that only authorized users have access to resources and information; and implementing technologies and processes that reduce the risk of malware. The cybersecurity CAP goal consisted of three priority areas with a total of nine performance indicators. Each of the nine performance indicators had an expected level of

[54] Each year, OMB and DHS work with an interagency group to develop the CIO FISMA metrics. These metrics are organized around the five cybersecurity framework core security functions and track agencies' progress in implementing cybersecurity capabilities.

[55] Although the CAP goal was in place for fiscal years 2015 through 2017, the scope of our review was for fiscal years 2016 and 2017.

performance, or target, for implementation. Table 3 shows the three priority areas and related performance indicators and targets of the cybersecurity CAP goal for fiscal years 2015 through 2017.

According to agency CIO assessments for fiscal year 2017, 6 of the 23 agencies met all 9 targets for the cybersecurity CAP goal. More specifically,

- 8 agencies met all four targets for the information security continuous monitoring priority area;
- 16 agencies met the two targets for the identity, credential, and access management priority area; and
- 17 agencies met all three targets for the anti-phishing and malware defense priority area.

In addition, CIOs reported that agencies were making progress in meeting the targets for the nine performance indicators for fiscal years 2016 and 2017, with increases in the number of agencies meeting the targets within each of the three priority areas.

However, although the number of agencies that met the targets in individual priority areas showed a net increase, not all agencies maintained their status. For example, the CIO for one agency reported meeting all three targets for the anti-phishing and malware defense priority area in fiscal year 2016, but reported that the agency only met two of the three targets in fiscal year 2017. Figure 5 shows the number of agencies that reported meeting each of the targets within the individual cybersecurity CAP goal priority areas for fiscal years 2016 and 2017.

Although the CIOs for only six agencies reported meeting each of the targets associated with all nine performance indicators for the three cybersecurity CAP goal priority areas, the CIOs at an additional eight agencies reported meeting each target for two of the three priority areas. Specifically,

- one CIO reported that its agency met each of the targets for the (1) information security continuous monitoring and (2) identity, credential, and access management priority areas;
- another CIO reported that its agency met each of the targets for the (1) information security continuous monitoring and (2) anti-phishing and malware defense priority areas; and
- the CIOs at six other agencies met each of the targets for the (1) identity, credential, and access management and (2) anti-phishing and malware defense priority areas.

Table 3. Priority Areas, Performance Indicators, and Targets for the Cybersecurity Cross-Agency Priority Goal, Fiscal Years 2015–2017

Priority area	Performance indicator	Target
Information security continuous monitoring is the provision that covers ongoing observation, assessment, analysis, and diagnosis of an organization's cybersecurity posture, hygiene, and operational readiness.	Hardware asset management	Implemented at 95 percent for the agency
	Software asset management	Implemented at 95 percent for the agency
	Vulnerability management	Implemented at 95 percent for the agency
	Secure configuration management	Implemented at 95 percent for the agency
Identity, credential, and access management is the implementation of a set of capabilities that are intended to ensure users must authenticate their identities in order to use	Implementation of personal identity verification for unprivileged users	Implemented at 85 percent for unprivileged users
	Implementation of personal identity	Implemented at 100 percent for privileged
information technology resources and have access to only those resources that are required for their job function.	verification for privileged users	users
Anti-phishing and malware defense is the implementation of technologies, processes, and training that are intended to reduce the risk of malware introduced through email and malicious or compromised web sites.	Anti-phishing defense	Implemented at 90 percent for 5 of the top 7 anti-phishing defenses
	Malware defenses	Implemented at 90 percent for 3 of the top 5 malware defenses
	Other defenses	Implemented at 90 percent for 2 of the top 4 other defenses

Source: GAO analysis of Office of Management and Budget Fiscal Year 2017 Federal Information Security Modernization Act of 2014 Annual Report to Congress. | GAO-19-105.

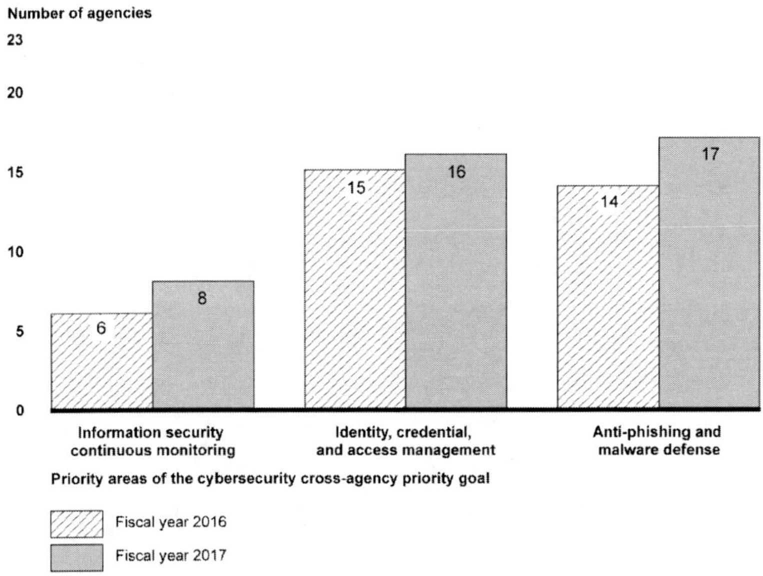

Source; GAO analysis of agency reported data. | GAO-19-105.

The 23 selected civilian agencies are the Departments of Agriculture, Commerce, Education, Energy, Health and Human Services, Homeland Security, Housing and Urban Development, the Interior, Justice, Labor, State, Transportation, the Treasury, and Veterans Affairs; the Environmental Protection Agency; General Services Administration; National Aeronautics and Space Administration; National Science Foundation; Nuclear Regulatory Commission; Office of Personnel Management; Small Business Administration; Social Security Administration; and the U.S. Agency for International Development.

Figure 5. Number of the 23 Selected Civilian Agencies That Reported Meeting Targets for the Cybersecurity Cross-Agency Priority Goal Priority Areas, Fiscal Years 2016 and 2017.

In fiscal year 2018, the President's Management Agenda[56] replaced the three cybersecurity-focused CAP goal priority areas with updated performance indicators, most of which are to be met by 2020:

---

[56] The President's Management Agenda is intended to lay out a long-term vision for modernizing the federal government in key areas that will improve the ability of agencies to deliver mission outcomes, provide excellent service, and effectively steward taxpayer dollars on behalf of the American people.

1) the manage asset security priority area is similar to the information security continuous monitoring priority area from the previous CAP goal and has a focus on understanding the assets and users on agency networks. In addition to hardware asset and software asset management, this priority area includes performance indicators for authorization and mobile device management.
2) the limit personnel access priority area focuses on issues of access management. This area includes performance indicators for using automated access management and managing access for privileged network and high-impact system users. The privileged network access management performance indicator is a continuation of the identity, credential, and access management priority area of the previous cybersecurity CAP goal. Therefore, agencies are expected to complete this metric by the end of the fiscal year 2018 FISMA reporting year.
3) the protect networks and data priority area, which is similar to the anti-phishing and malware defense priority area from the previous cybersecurity CAP goal, has three new performance indicators: intrusion detection and prevention, exfiltration and enhanced defenses, and data protection.

Appendix IV describes the updated cybersecurity-focused CAP priority areas and performance indicators in more detail.

OMB Determined That 13 of the 23 Civilian CFO Act Agencies Were Managing Cybersecurity Risk

In Executive Order 13800, the President directed OMB, in coordination with DHS, to assess and report to the executive branch on the sufficiency and appropriateness of federal agencies' processes for managing cybersecurity risks. For these risk management assessments,[57] OMB leveraged the FISMA metrics reported by agency CIOs and inspectors general for fiscal year 2017. The metrics addressed domains that correspond with the five core security functions identified in the cybersecurity framework. Table 4 lists these domains and their relationship to the core functions.

Table 4. *Federal Information Security Modernization Act* Reporting Metric Domains Leveraged by OMB to Assess Agency Risk Management Processes

Core security functions	Domains
Identify	Asset management and authorization Comprehensive risk management
Protect	Remote access protection Credentialing and authorization Network protection
Detect	Anti-phishing capabilities Malware defense capabilities Exfiltration and other capabilities
Respond	Planning and processes Evaluation and Improvement
Recover	Planning and testing personal impact process Back-up capacity

Source: GAO analysis of Office of Management and Budget Fiscal Year 2017 Federal Information Security Modernization Act of 2014 Annual Report to Congress. | GAO-19-105.

Based on OMB's evaluation of these domains, agency risk management processes related to the five core security functions and overall agency enterprise fell into one of the following three rating categories:

- managing risk: required cybersecurity policies, procedures, and tools are in use and the agency actively manages cybersecurity risks;

[57] Office of Management and Budget, *Federal Information Security Modernization Act of 2014, Annual Report to Congress Fiscal Year 2017* (Washington, D.C.: Mar. 2018).

- at risk: some essential policies, processes, and tools are in place to mitigate overall cybersecurity risk, but significant gaps remain; and
- high risk: key fundamental cybersecurity policies, processes, and tools are either not in place or not deployed sufficiently.

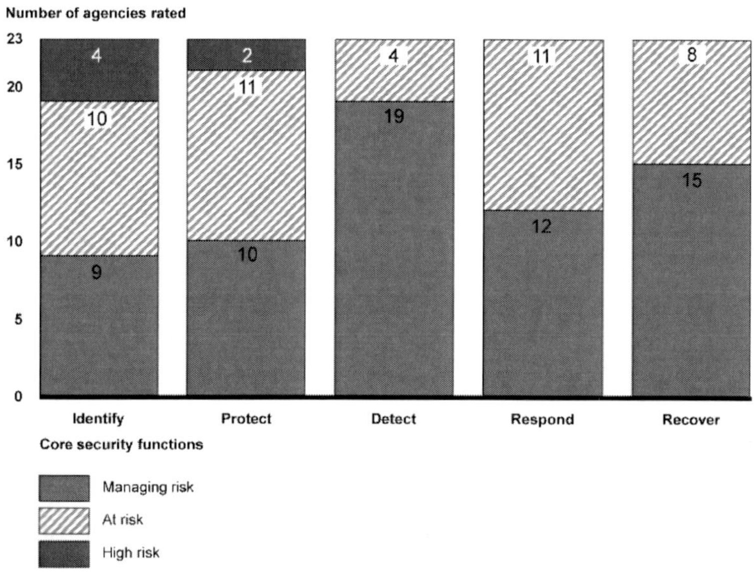

Source: GAO analysis of Office of Management and Budget Fiscal Year 2017 Federal Information Security Modernization Act of 2014 Annual Report to Congress. | GAO-19-105.

The 23 civilian agencies covered by the *Chief Financial Officers Act of 1990* are the Departments of Agriculture, Commerce, Education, Energy, Health and Human Services, Homeland Security, Housing and Urban Development, the Interior, Justice, Labor, State, Transportation, the Treasury, and Veterans Affairs; the Environmental Protection Agency; General Services Administration; National Aeronautics and Space Administration; National Science Foundation; Nuclear Regulatory Commission; Office of Personnel Management; Small Business Administration; Social Security Administration; and the U.S. Agency for International Development.

Figure 6. Risk Management Assessment Ratings by Core Security Function for the 23 Civilian *Chief Financial Officers Act of 1990* Agencies, Fiscal Year 2017.

For fiscal year 2017, OMB reported that not all agencies were managing risk. When considering each of the five core security functions, OMB reported that most of the 23 agencies were at risk or at high risk with regard to the *identify* and *protect* core security functions. Less than half of the 23 agencies were at risk with regard to the *detect*, *respond*, and *recover* core security functions. Overall, OMB determined that 13 agencies were managing risk and that the remaining 10 agencies were at risk of not effectively identifying, protecting, detecting, responding to, and if necessary, recovering from cyber intrusions. Figure 6 shows OMB's risk management assessment ratings by core security function across the 23 agencies for fiscal year 2017.

DHS AND OMB FACILITATED THE USE OF INTRUSION DETECTION AND PREVENTION CAPABILITIES TO SECURE FEDERAL AGENCY SYSTEMS, BUT FURTHER EFFORTS REMAIN

DHS and OMB, as required by law and policy, have taken various actions to facilitate the agencies' use of intrusion detection and prevention capabilities to secure federal systems. For example, DHS has developed an intrusion assessment plan, deployed NCPS to offer intrusion detection and prevention capabilities to agencies, and is providing tools and services to agencies to monitor their networks through its CDM program. However, NCPS still had limitations in detecting certain types of traffic and agencies were not sending all appropriate traffic through the system. Further, CDM was behind at meeting planned implementation dates, and agencies have requested additional training and guidance for these services. OMB has taken steps to improve upon agencies' capabilities, but has not completed a policy and strategy to do so, or fully reported on its assessment of agencies' capabilities.

DHS Has Taken Actions to Facilitate the Use of Intrusion Detection and Prevention Capabilities and to Make Improvements to Those Capabilities

The *Federal Cybersecurity Enhancement Act of 2015* requires DHS, in coordination with OMB, to develop and implement an intrusion assessment plan to proactively detect, identify, and remove intruders in agency information systems on a routine basis. The act also requires that the plan be updated, as necessary.

In December 2016, DHS documented its *Intrusion Assessment Plan*.[58] In the plan, DHS outlined tools, platforms, resources, and ongoing work that the department provides, and that are intended to help agencies detect, identify, and remove intruders on their networks and systems. The intrusion assessment plan also outlines a defense-in-depth strategy, which utilizes multiple layers of cybersecurity and deploys multiple capabilities in combination, to secure agencies' networks and information systems. For example, the plan calls for DHS to implement NCPS to provide a perimeter defense for the networks of federal civilian executive branch agencies, while the agencies are to deploy their own intrusion detection and prevention capabilities inside their networks. DHS submitted its intrusion assessment plan to OMB in January 2017.

DHS Has Worked to Improve NCPS, but Agencies Did Not Route All Traffic through Intrusion Detection and Prevention Capabilities Offered by this System

The *Federal Cybersecurity Enhancement Act of 2015* also requires DHS to deploy, operate, and maintain a capability to detect cybersecurity risks and prevent network traffic associated with such risks from transiting to or from an agency information system. In addition, the act requires that

[58] Department of Homeland Security, *Intrusion Assessment Plan: Fiscal Year 2016 Report to Congress* (Washington, D.C.: Dec. 22, 2016).

DHS make regular improvements to intrusion detection and prevention capabilities by deploying new technologies and modifying existing technologies. Further, the act requires agencies to use this capability on all information traveling between their information systems and any information system other than an agency information system.

DHS developed NCPS, operationally known as EINSTEIN, to provide the capabilities to detect and prevent potentially malicious network traffic from entering agency networks. Consistent with recommendations we made to DHS in January 2016, DHS has taken actions to improve these capabilities and has other actions underway.[59] For example, the department

- determined that enhancing NCPS's current intrusion detection approach to include functionality that would detect deviations from normal network behavior baselines would be feasible. In addition, according to DHS officials, the department was operationalizing functionality intended to identify malicious activity in network traffic otherwise missed by signature-based methods.
- determined that developing enhancements to current intrusion detection capabilities to facilitate the scanning of Internet Protocol Version 6 (IPv6)[60] traffic would be feasible. According to DHS officials, the department has developed plans to fully support IPv6 for several of its NCPS intrusion detection capabilities. Further, the department has developed implementation schedules and begun roll-out of the enhancements.

---

[59] GAO-16-294.

[60] In September 2010, the federal chief information officer issued a memorandum for agency chief information officers stating that the federal government is committed to the operational deployment and use of IPv6, and in July 2012, the Federal Chief Information Officer Council Strategy and Planning Committee issued a roadmap toward IPv6 adoption within the government. The roadmap stated that though both IPv4 (the legacy version of IP) and IPv6 are being used on the Internet, IPv4 is, by far, still the dominate protocol because of its legacy deployment. However, IPv6 traffic growth is inevitable due to the current state of IPv4 address exhaustion, creating an extreme supply and demand curve required to support communications between the U.S. government and its citizens and business partners worldwide.

- updated the tool it uses to manage and deploy intrusion detection signatures to include a mechanism to clearly link signatures to publicly available, open-source information.
- developed clearly defined requirements for detecting threats on agency internal networks and at cloud service providers to help better ensure effective support of information security activities. According to DHS officials, the department was also continuing pilot activities with cloud service providers to enhance protections of agency assets.
- developed processes and procedures for using vulnerability information, such as data from the CDM program as it becomes available, to help ensure the department is using a risk-based approach for the selection/development of future NCPS intrusion prevention capabilities.

Nevertheless, NCPS continues to have known limitations in its ability to identify potential threats. For example:

- NCPS does not have the ability to effectively detect intrusions across multiple types of traffic. Specifically, DHS determined that developing enhancements to current intrusion detection capabilities to facilitate the scanning of traffic related to supervisory control and data acquisition (SCADA)[61] control systems would not be feasible. However, according to DHS officials, the department is exploring capabilities that are intended to provide critical, cross-sector, real-time visibility into critical infrastructure companies that utilize SCADA systems. In addition, DHS determined that the scanning of encrypted traffic would not

[61] SCADA is one type of control system, which is a computer-based system used within many infrastructures and industries to monitor and control sensitive processes and physical functions. Control systems perform functions that range from simple to complex. They can be used to simply monitor processes—for example, the environmental conditions in a small office building—or to manage the complex activities of a municipal water system or a nuclear power plant. Control systems are vulnerable to cyberattack from inside and outside the control system network.

be feasible. Nevertheless, according to its officials, the department performed research on potential architectural, technical, and policy mitigation strategies that could provide both the protection and situational awareness for encrypted traffic. The department has actions under way to continue its research in this area.

- DHS does not always explicitly ask agencies for feedback or confirmation of receipt of NCPS-related notification. While the department had drafted a standard operating procedure related to its incident notification process, the policy did not instruct DHS analysts specifically to include a solicitation of feedback from agencies within the notification. Further, US-CERT could not provide any information regarding the timetable for when these procedures would take effect.
- Metrics for NCPS, as provided by DHS, do not provide information about how well the system is enhancing government information security or the quality, efficiency and accuracy of supporting actions. Without the deployment of comprehensive measures, DHS cannot appropriately articulate the value provided by NCPS. While the department had taken actions to develop new measures, these measures did not provide a qualitative or quantitative assessment of the system's ability to fulfill the system's objectives.
- NSD did not provide guidance to agencies on how to securely route their information to their Internet service providers. Without providing network routing guidance, NSD has no assurance that the traffic it sees constitutes all or only a subset of the traffic the customer agencies intend to send.

As shown in Table 5, as of October 2018, the department had implemented five of the nine recommendations and was in the process of implementing the remainder. However, until DHS completes implementation of the remaining recommendations, the effectiveness of NCPS's intrusion detection and prevention capabilities may be hindered.

Table 5. Status of GAO 2016 Recommendations to Improve the National Cybersecurity Protection System (NCPS) Capabilities as of October 2018[a]

Recommendation	Status	
	Imple-mented	In process
Determine the feasibility of enhancing NCPS's current intrusion detection approach to include functionality that would detect deviations from normal network behavior baselines.	✓	
Determine the feasibility of developing enhancements to current intrusion detection capabilities to facilitate the scanning of traffic not currently scanned by NCPS.	✓	
Update the tool the U.S. Computer Emergency Readiness Team uses to manage and deploy intrusion detection signatures to include the ability to more clearly link signatures to publicly available, open-source data repositories.	✓	
Consider the viability of using vulnerability information, such as data from the Continuous Diagnostics and Mitigation program as it becomes available, as an input into the development and management of intrusion detection signatures.		✓
Develop a timetable for finalizing the incident notification process, to ensure that customer agencies are being sent notifications of potential incidents, which clearly solicit feedback on the usefulness and timeliness of the notification.		✓
Develop metrics that clearly measure the effectiveness of NCPS's efforts, including the quality, efficiency, and accuracy of supporting actions related to detecting and preventing intrusions, providing analytic services, and sharing cyber-related information.		✓
Develop clearly defined requirements for detecting threats on agency internal networks and at cloud service providers to help better ensure effective support of information security activities.	✓	
Develop processes and procedures for using vulnerability information, such as data from the Continuous Diagnostics and Mitigation program as it becomes available, to help ensure DHS is using a risk-based approach for the selection/development of future NCPS intrusion prevention capabilities.	✓	
Work with customer agencies and the Internet service providers to document secure routing requirements in order to better ensure the complete, safe, and effective routing of information to NCPS sensors.		✓

Legend: ✓ A checkmark indicates that the recommendation was either implemented or in the process of being implemented, as designated by the column heading.

Source: GAO analysis of Department of Homeland Security data. | GAO-19-105.

[a]Recommendations are found in *Information Security: DHS Needs to Enhance Capabilities, Improve Planning, and Support Greater Adoption of its National Cybersecurity Protection System*. GAO-16-294 (Washington, D.C.: Jan. 28, 2016).

Table 6. Number of Civilian *Chief Financial Officers Act of 1990* Agencies[a] that Had Implemented the National Cybersecurity Protection System Capabilities through EINSTEIN, by Implementation Status, as of September 2017

Capability	Implemented	In progress	Deferred[b]	Not implemented
EINSTEIN 1 /	21	2	0	0
EINSTEIN 2				
EINSTEIN 3	23	0	0	0
Accelerated				
(DNS Sinkholing)[c]				
EINSTEIN 3	15	6	2	0
Accelerated				
(Email Filtering)[d]				

Source: Office of Management and Budget Federal Information Security Modernization Act of 2014 Annual Report to Congress Fiscal Year 2017. | GAO-19-105.

[a]These summary implementation data aggregate the status for parent agencies based on the lowest implementation status of all of their components.

[b]An agency with a deferred status faced a technical challenge to implement email filtering for its third-party, cloud-based email service. DHS continues to work with the affected agencies and their EINSTEIN 3 Accelerated service provider to engineer solutions.

[c]The Domain Name System (DNS) Sinkholing capability, used to detect and prevent malicious traffic from federal civilian networks, redirects malicious traffic to "safe servers," also known as "sinkhole" servers."

[d]The email filtering capability, used to protect against the use of email to deliver malware or induce users to download malware to infect agency networks, scans email destined for federal civilian networks for malicious attachments and other potential cyber threats before being delivered to end-users. According to DHS officials, as of November 2018, the completed number greatly increased since the September 2017 report, with the CFO Act agencies being at 99 percent completed.

In addition, the 23 civilian CFO Act agencies had implemented NCPS capabilities to varying degrees. In a March 2018 report, OMB reported that 21 (about 91 percent) of the 23 agencies had implemented the first two iterations of the NCPS capabilities.[62] In addition, 15 (about 65 percent) of the 23 agencies had implemented all three NCPS capabilities, as shown in Table 6 below.[63]

---

[62] Office of Management and Budget, *Federal Information Security Modernization Act of 2014, Annual Report to Congress Fiscal Year 2017* (Washington, D.C.: Mar. 2018).

[63] While the March 2018 report shows that DHS reported 17 of the 23 civilian CFO Act agencies implementing all three NCPS capabilities, two of the 17 agencies had not fully implemented

However, agencies did not route all network traffic for all information traveling between their information systems and any information system other than an agency information system through NCPS sensors.

For example, officials at 13 of 23 agencies stated that not all of their agency external network traffic flowed through NCPS.[64] To illustrate, officials at one agency estimated that 20 percent of their external network traffic did not flow through the system. In addition, 4 of the agencies in our review previously cited several challenges in routing all of their traffic through NCPS intrusion detection sensors, including capacity limitations of the sensors, agreements with external business partners that use direct network connections, interagency network connections that do not route through Internet gateways, use of encrypted communications mechanisms, and backup network circuits that are not used regularly. NSD officials stated that agencies are responsible for routing their traffic to the intrusion detection sensors, and DHS does not have a role in that aspect of NCPS implementation. As a result, potential cyberattacks may not be detected or prevented for a portion of the external traffic at federal agencies. As noted above, we previously recommended that DHS work with agencies to better ensure the complete routing of information to NCPS sensors.

DHS Has Taken Steps to Provide Advanced Network Security Tools, but Has Not Met Planned Implementation Dates

The *Federal Cybersecurity Enhancement Act of 2015* requires DHS to include, in the efforts of the department to continuously diagnose and mitigate cybersecurity risks, advanced network security tools to improve

---

the EINSTEIN 3 Accelerated Email Filtering capability. Accordingly, we have adjusted this number to reflect the number of agencies that have fully implemented all three capabilities.

[64] External network traffic (traffic that is routed through agency's external connections) must be routed through a Trusted Internet Connection. External connections include those connections between an agency's information system or network and the globally-addressable Internet or a remote information system or network and networks located on foreign territory.

the visibility of network activity and to detect and mitigate intrusions and anomalous activity.

According to DHS officials, the department is addressing the requirement to improve the visibility of network activity by including advanced network security tools as a part of CDM phase 3. In April 2018, we testified that DHS had previously planned to provide 97 percent of federal agencies with the services they needed for CDM phase 3 in fiscal year 2017.[65] In addition, according to OMB's annual FISMA report for fiscal year 2017, the CDM program was to continue to incorporate additional capabilities, including phase 3, in fiscal year 2018.[66]

However, DHS now expects initial operational capabilities[67] to be in place for phase 3 in fiscal year 2019. The department has awarded contracts of approximately $3.26 billion to support its Dynamic and Evolving Federal Enterprise Network Defense (also known as DEFEND) aspect of the CDM program, which is to include phase 3. DEFEND also is to provide coverage for existing agency deployments. According to DHS documentation, the task orders associated with DEFEND are to be issued between the second quarter of fiscal year 2018 and the second quarter of fiscal year 2024.

Agencies Indicated the Need for Additional Training and Guidance Related to NCPS and CDM

FISMA requires that DHS provide operational and technical assistance to agencies in implementing policies, principles, standards, and guidelines on information security. Toward this end, DHS has available training and guidance related to the implementation of the capabilities of NCPS (i.e., EINSTEIN) and CDM. Specifically:

---

[65] GAO, *Cybersecurity: DHS Needs to Enhance Efforts to Improve and Promote the Security of Federal and Private-Sector Networks*, GAO-18-520T (Washington, D.C.: Apr. 24, 2018).

[66] Office of Management and Budget, *Federal Information Security Modernization Act of 2014, Annual Report to Congress Fiscal Year 2017* (Washington, D.C.: Mar. 2018).

[67] Initial operational capability is a DHS-defined project milestone that is attained when the capabilities for a phase have been fully deployed to at least five agencies.

- According the DHS officials, the department offers training and guidance to agencies on EINSTEIN 1 implementation. For example, DHS established a program in which the Software Engineering Institute will provide training and mentoring to agencies looking to enhance their understanding of, and proficiency with, the EINSTEIN 1 capability (e.g., network traffic information).[68] NCPS program officials stated that agencies can use this service, which is available at no charge to them, on an unlimited basis as long as the requests relate to EINSTEIN 1. According to the officials, training and guidance related to EINSTEIN 2 and EINSTEIN 3 Accelerated is limited because DHS intentionally restricts the amount of data provided to agencies.
- According to DHS officials, the department also offers training and guidance to assist agencies with the implementation and use of resources associated with the CDM program, including webinars, guides, and computer-based training. The DEFEND contracts that the department awarded also include a mechanism for agencies to procure specialized tailored training, such as on the use of CDM tools. The department also offers customer advisory forums every other month that agencies are invited to attend. According to CDM program officials, the program's governance, among other topics, is commonly discussed during these forums. Further, the department provides agencies with guidance, such as various governance documents, best practices, and frequently asked questions, through a web portal that is made available by OMB. In addition, US-CERT offers the CDM training program, which is to provide CDM implementation resources.

[68] The Software Engineering Institute is a federally funded research and development center at Carnegie Mellon University. The institute works with DHS to strengthen the nation's resistance to cyber threats and improve the practice of cybersecurity.

Nevertheless, most agencies told us that they wanted DHS to provide more training and guidance as it relates to their implementation of the capabilities made available by NCPS and CDM. Specifically,

- Officials from 16 of 23 agencies reported that they wanted to receive additional training on NCPS capabilities. For example, officials at 5 agencies stated that they would like to receive training related to using network traffic information, understanding alerts, or implementing capabilities for cloud services. The officials also stated that they wanted training specific to agency security personnel.
- Officials from 19 of 23 agencies stated that they wanted to receive additional guidance related to NCPS's capabilities, but not all of the 19 provided specific details. For example, officials from at least 3 agencies stated that they wanted additional guidance such as, "how to" documents, descriptions of architecture details, or guidance documents that explain NCPS's capabilities so that agencies can gauge the gap between the security that the system provides and the security being provided by their own agency's capabilities.
- Officials from 21 of 23 agencies reported that they wanted to receive additional training on implementing CDM at their agencies. For example, officials from 7 agencies suggested that additional training on the use of the tools would be beneficial.
- Officials from 22 of 23 agencies stated that they wanted additional guidance as it relates to CDM implementation. For example, officials from one agency stated that they would like examples of best practices and successful deployments.

These requests for additional training and guidance demonstrate that agencies are either unaware of the available training and guidance, or that the training may not meet their needs. Until DHS coordinates with agencies to determine if additional training and guidance are needed, agencies may not be able to fully realize the benefits of the capabilities provided by the NCPS and CDM programs.

OMB Took Actions to Oversee Agency Implementation of Intrusion Detection and Prevention Capabilities and Report to Congress, but Did Not Fully Complete Required Actions

Although OMB took steps to report on agencies' implementation of intrusion detection and prevention capabilities, it did not report on all required actions. For example, the office did not submit DHS's intrusion plan to Congress as required by the *Federal Cybersecurity Enhancement Act of 2015*. In addition, OMB provided various reports to Congress that described agencies' intrusion detection and prevention capabilities, but the reports did not always include all information required by the act. Further, OMB developed a draft policy and strategy that were intended to improve agency capabilities, but it had not finalized these documents.

OMB Did Not Submit the Intrusion Assessment Plan to Congress or Fully Describe the Plan's Implementation in Other Reports

The *Federal Cybersecurity Enhancement Act of 2015* requires OMB to submit the intrusion assessment plan developed by DHS to the appropriate congressional committees no later than 6 months after the date of enactment of the act. The act also required OMB to submit to Congress a description of the implementation of the intrusion assessment plan and the findings of the intrusion assessments[69] conducted pursuant to the intrusion assessment plan no later than 1 year after the date of enactment of the act, and annually thereafter.

Although DHS developed and documented an intrusion assessment plan, which described a defense-in-depth approach to security, OMB did not submit the plan to Congress, as called for in the act. Even though DHS

[69] Intrusion assessments, as defined by the *Federal Cybersecurity Enhancement Act of 2015*, refer to actions taken under the intrusion assessment plan to identify and remove intruders in agency information systems.

submitted the plan to OMB in January 2017, OMB had not submitted it to Congress as of October 2018 (21 months after DHS submitted the plan and 28 months past the due date).

On the other hand, OMB did submit its own reports to Congress which generally described elements of the implementation of DHS's intrusion assessment plan and intrusion assessment findings. In September 2017, OMB issued its analysis of agencies' implementation of intrusion detection and prevention capabilities, or more specifically, agencies' implementation of the various versions of NCPS.[70] In addition, the office's annual FISMA report, issued most recently in March 2018,[71] generally covered elements of the intrusion assessment plan. OMB personnel within the Office of the Federal CIO believed that these two reports, along with a process the office had initiated to validate incidents across the government, addressed the requirement for OMB to submit to Congress a description of the implementation of the intrusion assessment plan and the findings of the intrusion assessments conducted pursuant to the plan.

However, the September 2017 and March 2018 reports did not address other elements described in DHS's intrusion assessment plan. For example, OMB did not describe agency roles associated with segmenting their networks, identifying key servers based on threat and impact, ensuring all applications are appropriately tracked and configured, and categorizing and tagging data based on threat and impact. While OMB has provided important information to congressional stakeholders through its own reports, until it submits the plan and addresses all elements described in DHS's intrusion assessment plan, it will continue to be remiss in providing timely and sufficiently detailed information regarding the intrusion assessment plan to congressional stakeholders to support their oversight responsibilities.

---

[70] Office of Management and Budget, *Agency Application of the National Cybersecurity Protection System and Intrusion Detection and Prevention Capabilities* (Washington, D.C.: September 2017).

[71] Office of Management and Budget, *Federal Information Security Modernization Act of 2014, Annual Report to Congress Fiscal Year 2017* (Washington, D.C.: Mar. 2018).

OMB Submitted Its Analysis of Agencies' Application of Intrusion Detection and Prevention Capabilities, but Did Not Include the Degree to Which the Capabilities Had Been Applied

The *Federal Cybersecurity Enhancement Act of 2015* also required that OMB submit an analysis of agencies' application of the intrusion detection and prevention capabilities to Congress no later than 18 months after the date of enactment of the act, and annually thereafter. OMB was to include a list of federal agencies and the degree to which each agency had applied the intrusion detection and prevention capabilities in this analysis.

As discussed previously in this chapter, OMB issued its analysis of agencies' implementation of intrusion detection and prevention capabilities in September 2017. However, the analysis did not include the degree to which agencies had applied the intrusion detection and prevention capabilities. For example, the analysis did not reflect that not all agencies were using this capability on all information traveling between their systems and any system other than an agency system, as required by the act. Until OMB includes the degree to which agencies have applied intrusion detection and prevention capabilities in its analysis, it cannot provide congressional stakeholders with an accurate portrayal of the extent to which the capabilities are detecting and preventing potential intrusions.

The Federal Chief Information Officer Reported on Intrusion Detection and Prevention Capabilities, but Did Not Address All Elements Required by the Federal Cybersecurity Enhancement Act of 2015

The *Federal Cybersecurity Enhancement Act of 2015* further required that the Federal Chief Information Officer, within OMB, submit a report to Congress no earlier than 18 months after the date of enactment, but no later than 2 years after that date, assessing the intrusion detection and intrusion prevention capabilities that DHS made available to agencies. The act required that the report address (1) the effectiveness of DHS's system used for detecting, disrupting, and preventing cyber-threat actors, including advanced persistent threats, from accessing agency information and agency information systems; (2) whether the intrusion detection and prevention capabilities, continuous diagnostics and mitigation, and other systems deployed are effective in securing federal information systems; (3) the costs and benefits of the intrusion detection and prevention capabilities, including as compared to commercial technologies and tools, and including the value of classified cyber threat indicators; and (4) the capability of agencies to protect sensitive cyber threat indicators and defensive measures if they were shared through unclassified mechanisms for use in commercial technologies and tools.

In a report issued in September 2018 (about 8 months past the required due date), the Federal Chief Information Officer provided Congress an assessment of intrusion detection and intrusion prevention capabilities across the federal enterprise. The report pointed out, among other things, that agencies did not possess or properly deploy capabilities to detect or prevent intrusions or minimize the impact of intrusions when they occur. In addition, the report acknowledged the need to improve the effectiveness of intrusion detection and intrusion prevention capabilities and stated that OMB would track performance through the CAP goal and annual FISMA reports.

However, the report did not address all of the requirements specified in the act. For example, the report did not address whether DHS's system (i.e., NCPS) was effective in detecting advanced persistent threats. In addition, the report did not include a comparison of the costs and benefits of the intrusion detection and prevention capabilities versus commercial technologies and tools, or the value of classified cyber threat indicators. Further, the report did not address the capability of agencies to protect sensitive cyber threat indicators and defensive measures. Until OMB updates the Federal CIO report to address all of the requirements specified in the act, it will continue to be remiss in providing timely and sufficiently detailed information, such as that related to costs and benefits, among other elements in the act, to congressional stakeholders to support their oversight responsibilities.

OMB Initiated Plans for Improving Agencies' Implementation of Intrusion Detection and Prevention Capabilities, but Has Not Completed a Policy and Strategy

In addition to OMB's responsibilities in the *Federal Cybersecurity Enhancement Act of 2015*, OMB has initiated plans for further improving agencies' intrusion detection and prevention capabilities. In response to a tasking in Executive Order 13800, the Director of the American Technology Council coordinated the development of a report to the President from the Secretary of DHS, the Director of OMB, and the Administrator of the General Services Administration, regarding the modernization of federal information technology (IT).

The report, *Report to the President on Federal IT Modernization*, identified actions that OMB should take for (1) prioritizing the modernization of high-risk, high-value assets and (2) modernizing the Trusted Internet Connection (TIC) program[72] and NCPS to improve protections,[73] remove barriers, and enable commercial cloud migration. For example, OMB was to take the following actions subsequent to the December 13, 2017 report issuance date:

- Within 60 days: Update a TIC policy to address challenges with agencies' perimeter-based architectures, such as the modernization of NCPS. In addition, introduce a "90 day sprint" during which approved projects would pilot proposed changes in TIC requirements.

---

[72] In November 2007, OMB issued M-08-05 that announced the Trusted Internet Connections Initiative, which is intended to improve the federal government's security posture by reducing and consolidating external network connections, including Internet connections, currently in use by the government, and by centrally monitoring the traffic passing through these connections for potentially malicious activity. All federal agencies in the executive branch, except for the Department of Defense, are required to implement the initiative. Although the initiative is intended to secure connections to the Internet, other external connections to potentially unsecured systems must also be routed through an approved TIC access point, even if they do not pass through the Internet.

[73] The modernization report notes that TIC gateways apply common security protections for agencies, as well as common intrusion detection, information sharing, and prevention capabilities under DHS's NCPS.

- Within 90 days: Update the annual FISMA and CAP goal metrics to focus on those critical capabilities that were most commonly lacking among agencies and focus oversight assessments on high-value assets.
- Within 120 days: In conjunction with DHS, develop a strategy for optimally realigning resources across agencies to reduce the risk to high-value assets and respond to cybersecurity incidents for those assets.

OMB has taken steps toward implementing several, but not all, of these actions. For example, it introduced a "90 day sprint" and, according to knowledgeable OMB staff, the outcomes of this action are directly informing changes in TIC requirements. In addition, OMB updated the annual FISMA and CAP goal metrics by including several metrics that focus on high-value assets. The updated FISMA and CAP goal metrics went into effect in April 2018.

However, while OMB had taken steps toward updating the TIC policy and developing a strategy for optimally realigning resources, the policy and strategy were in draft and had not yet been finalized as of October 2018. The agency did not specify a time frame for finalizing the policy and strategy. Until OMB finalizes the TIC policy and the strategy for optimally realigning resources, the enhancements offered through the policy and strategy are unlikely to be realized.

SELECTED AGENCIES HAD NOT CONSISTENTLY IMPLEMENTED CAPABILITIES TO DETECT AND PREVENT INTRUSIONS

FISMA requires agencies to provide information security protections to prevent unauthorized access to their systems and information. Officials from the 23 selected agencies reported to us that they generally took steps to meet this requirement by augmenting the tools and services provided by

DHS with their own intrusion detection and prevention capabilities. However, agencies did not consistently implement five key capabilities specified by DHS and NIST guidance. In addition, most of the agencies did not fully implement any of the phases of DHS's CDM program that is intended to improve their capabilities to detect and prevent intrusions.

Few Agencies Had Fully Implemented Required Email Protections

Binding Operational Directive (BOD) 18-01 instructs agencies to enhance email security.[74] These enhancements include enabling encrypted email transmission, ensuring that receiving mail servers know what actions the agency would like taken when an email falsely claims to have originated from the agency, and removing certain insecure protocols, among others.[75] The final deadline for implementing all BOD 18-01 requirements was October 16, 2018. Additionally, NIST SP 800-53 Revision 4 recommends that security awareness training include training on how to recognize and prevent spear-phishing attempts.

As of September 2018, only 2 of the 23 agencies reported implementing all of the email requirements. For the remaining 21 agencies:

- 9 agencies stated that their agency had plans to implement all enhancements by the October 2018 deadline,
- 1 agency was uncertain whether it would meet the deadline, and
- 11 stated they would not be able to meet the deadline.[76]

---

[74] According to OMB, email, by way of phishing attacks, remains one of the most common threat vectors (or avenues of attack) across the government. See *Federal Cybersecurity Risk Determination Report and Action Plan* (Washington, D.C.: May 2018).

[75] Department of Homeland Security, *Enhance Email and Web Security*, BOD 18-01 (Washington, D.C.: Oct. 16, 2017).

[76] In September 2018, DHS issued a temporary policy exception notice to federal agencies for BOD 18-01's weak email cipher requirement. The notice requested that agencies impacted by the notice submit to DHS preventive security measures and a mitigation strategy that would be in place until these matters would be resolved. Evaluating DHS's temporary policy exception notice and agency actions taken as a result of the notice was not within the scope of this review. Of the 11 agencies that said they would not meet the October 2018

By contrast, the majority of agencies (22 of 23) reported that they had trained staff on spear-phishing exercises, as recommended by NIST SP 800-53 Revision 4. Officials at the remaining agency told us that the agency planned to have spear-phishing exercises in fiscal year 2019. Such training should help ensure that phishing will be a less effective attack vector against the majority of agencies. While agencies benefit from secure protocols and spear-phishing training, implementing the remaining BOD 18-01 email requirements would provide additional protection to agency information systems.

Agencies Informed GAO That They Often Had Not Implemented Four Key Capabilities

NIST recommends that federal agencies deploy intrusion detection and prevention capabilities. These capabilities include monitoring cloud services, using host-based intrusion prevention systems, monitoring external and internal network traffic, and using a security information and event management (SIEM) system. However, in our semi-structured interviews of the 23 agencies, officials told us that they often had not implemented many of these capabilities. Such inconsistent implementation exposes federal systems and the information they contain to additional risk. As part of their continuing oversight efforts, OMB and DHS can use the information below to work with agencies to identify obstacles and impediments affecting the agencies' abilities to implement these capabilities.

Less Than Half of the Selected Agencies That Used Cloud Services Monitored Their Cloud-Related Traffic

NIST SP 800-53 Revision 4 states that agencies should monitor and control communications at the external boundary of the network. However,

deadline, 4 stated that they could not meet the weak email cipher requirement due to their reliance on external email vendors. Two of the 4 said they were unable to meet other requirements as well.

as of June 2018, fewer than half of the agencies that used cloud computing services were monitoring cloud traffic. Specifically:

- 10 of 22 agencies that used Infrastructure as a Service were monitoring inbound and outbound Infrastructure as a Service traffic,
- 7 of 21 agencies that used Platform as a Service were monitoring inbound and outbound Platform as a Service traffic, and
- 10 of 23 agencies that used Software as a Service were monitoring inbound and outbound Software as a Service traffic.

Without monitoring traffic to and from cloud service providers, agencies risk a greater chance of malicious cloud activity detrimentally affecting agency information security.

Several Selected Agencies Had Not Fully Deployed Host-Based Capabilities

NIST SP 800-53 Revision 4 states that agency internal monitoring may be achieved by utilizing intrusion prevention capabilities. These capabilities include using host-based intrusion prevention systems to provide defense at an individual system or device level by protecting against malicious activities. Host-based capabilities include memory-based protection[77] and application whitelisting.[78]

As of June 2018, officials at the 23 agencies reported the following to us:

- 16 agencies used host-based intrusion prevention capabilities,
- 15 agencies used memory-based protection, and
- 8 agencies used host-based application whitelisting.

---

[77] Memory-based protections are safeguards that protect memory from unauthorized code execution.

[78] An application whitelist is a list of applications and application components that an agency has authorized for use on its hosts.

Until host-based intrusion protections are fully deployed, agencies will be at greater risk of malicious activity adversely affecting agency operations.

Not All Selected Agencies Monitored External and Internal Traffic

NIST SP 800-53 Revision 4 also states that agencies should monitor and control communications at the external boundary of the network and at key internal boundaries (e.g., network traffic). NIST guidance also stated that an agency should deploy monitoring devices strategically within the network to detect essential information.

However, the agencies in our review did not always monitor external and internal traffic. For example, of the 23 agencies:

- 5 reported that they were not monitoring inbound or outbound direct connections to outside entities.
- 11 reported that they were not persistently monitoring inbound encrypted traffic.
- 8 reported that they were not persistently monitoring outbound encrypted traffic.

In addition, 13 agencies reported they were not using a network-based session capture solution.[79] Of the 10 agencies that reported using this solution, officials from 2 agencies stated that they were not capturing workstation-to-workstation connections. Without thorough monitoring of external and internal traffic, agencies will have less assurance that they are aware of compromised or potentially compromised traffic within their network.

[79] A network-based session capture solution records information exchanged across an agency's network.

Most Agencies Reported Using a Security Information and Event Management Capability, but Did Not Always Use this Capability to Analyze Potential Threats

NIST SP 800-53 Revision 4 states that agencies should establish enhanced monitoring capabilities. Such capabilities should include automated mechanisms that collect and analyze incident data for increased threat and situational awareness. According to NIST, a security information and event management (SIEM) system analyzes data from different sources and identifies and prioritizes significant events. Sources of data used by SIEM systems include logs from database systems, network devices, security systems, web applications, and workstation operating systems.[80]

Of the 23 agencies that we reviewed, 21 reported using a SIEM capability. Over half of the agencies employing a SIEM used one or more of their logs to match against known vulnerabilities and advanced persistent threats, as well as to create real-time alerts. For example, of the 21 agencies:

- 14 agencies reported collecting database logs, but only 7 agencies reported using the logs to match against known vulnerabilities and persistent threats and to create real-time alerts;
- 20 agencies reported collecting network logs, but only 13 agencies reported using them to match against known vulnerabilities and persistent threats and to create real-time alerts;
- 21 agencies reported collecting security logs, but only 13 reported using them to match against known vulnerabilities and persistent threats and to create real-time alerts;
- 15 agencies reported collecting web application logs, but only 9 agencies reported using them to match against known vulnerabilities and persistent threats and to create real-time alerts; and

[80] According to NIST, logs are records of the events occurring within an organization's systems and networks. Logs from security software, operating systems, and applications typically contain information that includes security-related data.

- 13 agencies reported collecting workstation logs, but only 8 agencies reported using them to match against known vulnerabilities and persistent threats and to create real-time alerts.
- Only 5 agencies collected all 5 types of logs and used them to match against known vulnerabilities and persistent threats and to create real-time alerts.

By not fully using SIEM capabilities, agencies will have less assurance that relevant personnel will be aware of possible weaknesses or intrusions.

Agencies Are in the Process of Implementing DHS' CDM Program, but Most Agencies Have Not Fully Implemented Any of the Program Phases

To further enhance their intrusion detection and prevention capabilities, the 23 civilian CFO Act agencies were in the process of implementing DHS's CDM program. As previously discussed, Phase 1 of the program involves deploying products to automate hardware and software asset management, configuration settings, and common vulnerability management capabilities. Phase 2 intends to address privilege management and infrastructure integrity by allowing agencies to monitor users on their networks and to detect whether users are engaging in unauthorized activity. Phase 3 is intended to assess agency network activity and identify any anomalies that may indicate a cybersecurity compromise.

As of June 2018, most agencies had not fully implemented any of the three phases. As shown in Figure 7, 15 agencies had partially implemented phase 1, 21 had partially or not yet begun to implement phase 2, and none of the agencies had fully implemented phase 3.

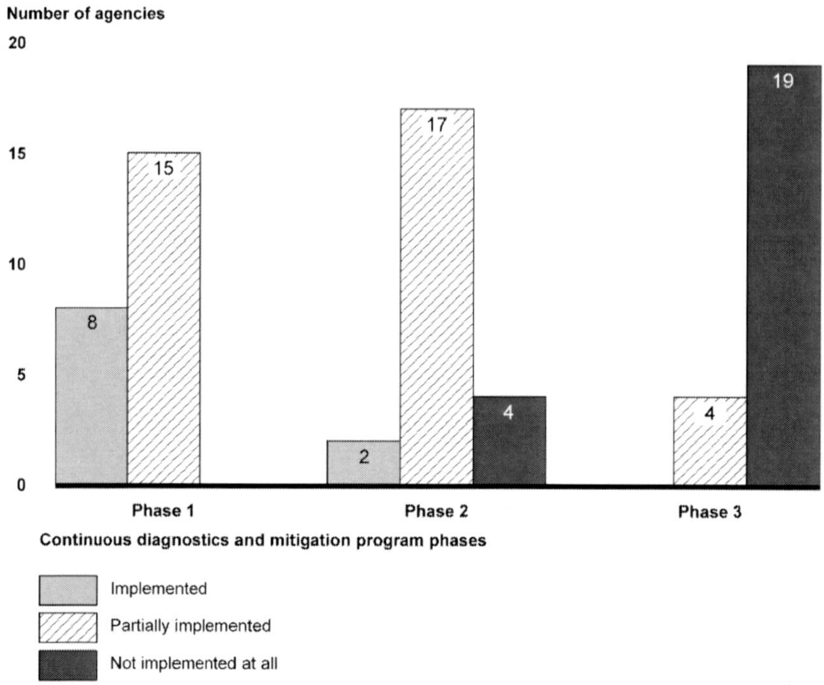

Source: GAO analysis based on agency provided data. | GAO-19-105.

Figure 7. Civilian *Chief Financial Officers Act of 1990* Agencies' Implementation of DHS's Continuous Diagnostics and Mitigation Program Phases, as of June 2018.

Agencies' implementation status has been affected, at least in part, due to delays in DHS's deployment of the program phases. As a result, federal systems will remain at risk until the program is fully deployed.

CONCLUSION

Many agencies have not effectively implemented the federal approach and strategy for securing information systems. For example, the inspectors general for 17 of the 23 selected agencies reported that their agencies had not effectively implemented their information security programs and had significant information security deficiencies associated with internal control over financial reporting. In addition, CIOs for 17 agencies reported not meeting all nine targets for the cybersecurity cross-agency priority goal. Further, OMB determined that that only 13 of the 23 agencies were managing risks to their overall enterprise, while the other 10 agencies were at risk. Until agencies more effectively implement the government's approach and strategy, federal systems will remain at risk.

DHS and OMB have initiatives underway that are intended to further improve agencies' security posture. However, although DHS had provided training and guidance for NCPS and CDM, agencies expressed the need for more. In addition, OMB had also not finalized its policy and strategy aimed at addressing challenges with perimeter security and protecting high value assets, respectively. OMB had also not provided useful information to Congress, such as a description of agencies' implementation of DHS's intrusion assessment plan, the degree to which agencies are using NCPS, a complete analysis of agencies' implementation of DHS's intrusion assessment plan, or the costs and benefits of using commercial tools.

Although agencies' officials reported various efforts underway to enhance their agency's intrusion detection and prevention capabilities, implementation efforts across the federal government were not consistent. OMB and DHS can use the information provided in this chapter to work with agencies to identify obstacles and impediments affecting the agencies' abilities to implement these capabilities.

RECOMMENDATIONS FOR EXECUTIVE ACTION

We are making a total of nine recommendations, including two to DHS and seven to OMB. Specifically:

- The Secretary of DHS should direct the Network Security Deployment division to coordinate further with federal agencies to identify training and guidance needs for implementing NCPS and CDM. (Recommendation 1)

- The Secretary of DHS should direct the appropriate staff to work with OMB to follow up with agencies to identify obstacles and impediments affecting their abilities to implement intrusion detection and prevention capabilities. (Recommendation 2)
- The Director of OMB should submit the intrusion assessment plan to the appropriate congressional committees. (Recommendation 3)
- The Director of OMB should report on implementation of the defense-in-depth strategy described in the intrusion assessment plan, including all elements described in the plan. (Recommendation 4)
- The Director of OMB should update the analysis of agencies' intrusion detection and prevention capabilities to include the degree to which agencies are using NCPS. (Recommendation 5)
- The Director of OMB should direct the Federal CIO to update her report to Congress to include required information, such as detecting advanced persistent threats, a comparison of the costs and benefits of the capabilities versus commercial technologies and tools, and the capability of agencies to protect sensitive cyber threat indicators and defense measures. (Recommendation 6)
- The Director of OMB should establish a time frame for finalizing the Trusted Internet Connections policy intended to address challenges with agencies' perimeter-based architectures and issue it when finalized. (Recommendation 7)
- The Director of OMB should establish a time frame for finalizing the strategy for realigning resources across agencies to protect high-value assets and issue it when finalized. (Recommendation 8)
- The Director of OMB should direct the Federal CIO to work with DHS to follow-up with agencies to identify obstacles and impediments affecting their abilities to implement intrusion detection and prevention capabilities. (Recommendation 9)

AGENCY COMMENTS AND OUR EVALUATION

We provided a draft of this chapter to OMB and the 23 civilian CFO Act agencies, including DHS, covered by our review. In response, OMB provided comments via email, and DHS and three other agencies (the Department of Commerce, Social Security Administration, and U.S. Agency for International Development) provided written comments, which are reprinted in appendices V through VIII, respectively. The 19 remaining agencies (the Departments of Agriculture, Education, Energy, Health and Human Services, Housing and Urban Development, the Interior, Justice, Labor, State, Transportation, the Treasury, and Veterans Affairs; as well as the Environmental Protection Agency, General Services Administration, National Aeronautics and Space Administration, National Science Foundation, Nuclear Regulatory Commission, Office of Personnel Management, and Small Business Administration) stated via email that they had no comments.

In its comments, which the OMB liaison provided to GAO via email on December 7, 2018, OMB did not state whether it agreed or disagreed with the seven recommendations that we made to it. Rather, according to the liaison, OMB agreed with the facts in our draft report, but found that the report did not reflect the agency's rationale for not submitting the DHS intrusion assessment plan to Congress and a report on the implementation of the plan, as required by the *Federal Cybersecurity Enhancement Act of 2015*. The liaison stated that OMB is working closely with DHS to provide strategic direction in assessing gaps in, and modernizing, the manner in which intrusion detection and prevention capabilities are delivered to the federal government. Further, in a subsequent email on December 10, 2018, OMB said it believes the Federal CIO's September 2018 report to Congress, along with data provided in OMB's fiscal year 2017 FISMA report to Congress, achieves the outcomes sought by the *Federal Cybersecurity Enhancement Act of 2015* and demonstrates OMB's continuous engagement with DHS across the evolution of the intrusion detection and prevention program.

As stated in our report, we acknowledge that OMB has provided important information to congressional stakeholders through its reports. However, OMB's reports did not cover all outcomes described in the act. For example, as we pointed out, these reports did not fully address implementation of the defense-in-depth strategy described in DHS's intrusion assessment plan. In addition, although OMB reported on several elements required by the *Federal Cybersecurity Enhancement Act of 2015*, it did not report on all of the required elements. For example, the reports did not address whether DHS's NCPS was effective in detecting advanced persistent threats. The reports also did not include a comparison of the costs and benefits of the intrusion detection and prevention capabilities versus commercial technologies and tools, or the value of classified cyber threat indicators. Further, the reports did not address the capability of agencies to protect sensitive cyber threat indicators and defensive measures. Accordingly, we maintain that our recommendations for OMB to report on required elements in the *Federal Cybersecurity Enhancement Act of 2015* are warranted.

In addition, OMB suggested that we revise our recommendations to the agency to include a shared responsibility with DHS to help drive desired outcomes. However, six of the seven recommendations we are making to OMB are related to specific OMB responsibilities cited in either the *Federal Cybersecurity Enhancement Act of 2015* or the *Report to the President on Federal IT Modernization*. As such, we believe the recommendations are appropriately addressed to OMB. Furthermore, our recommendations do not prevent OMB from working with DHS to implement them. Our seventh recommendation to OMB—to work with DHS to follow up with agencies to identify obstacles and impediments affecting their abilities to implement intrusion detection and prevention capabilities—includes a shared responsibility with DHS. OMB also provided technical comments, which we incorporated into the report, as appropriate.

Subsequent to providing initial comments on our draft report, OMB issued a memorandum intended to provide a strategy for realigning resources across agencies to protect high-value assets.[81] This action addresses our recommendation 8, which called for the Director of OMB to establish a time frame for finalizing the strategy for realigning resources across agencies to protect high-value assets, and to issue the strategy when finalized.

In its comments, DHS stated that it concurred with the two recommendations we made to the department. DHS stated that it expects to implement the recommendations in 2019.

The Department of Commerce commented that the report was reasonable and that the department agreed with the findings and recommendations.

In its comments, the Social Security Administration stated that protecting its networks and information is a critical priority. According to the agency, it continued to make improvements in fiscal year 2018, such as improvements and progress in securing applications, leveraging the cloud, managing its assets and vulnerabilities, strengthening its network and incident response capabilities, improving its security training, and enhancing the overall effectiveness of its cybersecurity program.

Finally, the U.S. Agency for International Development commented that its inspector general had improved the agency's capability maturity ratings for core security functions in fiscal year 2018. The agency also pointed out that it was the only selected agency in which fiscal year 2017 indicators of effectiveness in implementing the federal approach and strategy for securing information systems were all positive (as noted in Appendix III).

We are sending copies of this chapter to appropriate congressional committees, the Director of OMB, the heads of the 23 civilian CFO Act agencies and their inspectors general, and other interested congressional parties.

Gregory C. Wilshusen
Director, Information Security Issues

---

[81] Office of Management and Budget, *Strengthening the Cybersecurity of Federal Agencies by Enhancing the High Value Asset Program*, M-19-03 (Washington, D.C.: Dec. 10, 2018).

APPENDIX I: OBJECTIVES, SCOPE, AND METHODOLOGY

The *Federal Cybersecurity Enhancement Act of 2015*, which was enacted December 18, 2015, included a provision for GAO to report on the effectiveness of the federal government's approach and strategy for securing agency information systems, including intrusion detection and prevention capabilities.[82] The objectives of our review were to assess: (1) the reported effectiveness of selected agencies' implementation of the federal government's approach and strategy to securing agency information systems; (2) the extent to which the Office of Management and Budget (OMB) and the Department of Homeland Security (DHS) have facilitated the use of intrusion detection and prevention capabilities to secure federal agency information systems; and (3) the extent to which selected agencies reported implementing intrusion detection and prevention capabilities.

Selected agencies for our review were the 23 civilian agencies covered by the *Chief Financial Officers Act of 1990* (CFO Act).[83] We did not include the Department of Defense because the *Federal Cybersecurity Enhancement Act of 2015* only pertains to civilian agencies. Because we focused our work on the 23 civilian agencies, results from these reviews are not generalizable to the entire federal government.

To assess the reported effectiveness of agencies' implementation of the federal government's approach and strategy to securing agency information systems, we

---

[82] The act is a part of the *Consolidated Appropriations Act*, 2016, Pub. L. No. 114-113, div. N, title II, subtitle B, 129 Stat. 2242, 2963 -2975 (Dec. 18, 2015).

[83] The 23 civilian *Chief Financial Officers Act of 1990* agencies are the Departments of Agriculture, Commerce, Education, Energy, Health and Human Services, Homeland Security, Housing and Urban Development, the Interior, Justice, Labor, State, Transportation, the Treasury, and Veterans Affairs; the Environmental Protection Agency; General Services Administration; National Aeronautics and Space Administration; National Science Foundation; Nuclear Regulatory Commission; Office of Personnel Management; Small Business Administration; Social Security Administration; and the U.S. Agency for International Development.

- described the federal government's approach and strategy by summarizing the Federal Information Security Modernization Act of 2014 (FISMA),[84] Executive Order 13800, Strengthening the Cybersecurity of Federal Networks and Critical Infrastructure,[85] and the National Institute of Standards and Technology's (NIST) Framework for Improving Critical Infrastructure Cybersecurity[86] (cybersecurity framework).
- assessed the reported effectiveness of agencies' implementation of the approach and strategy by reviewing annual reports from OMB and the inspectors general (IG) of the 23 civilian CFO Act agencies regarding the reported implementation of FISMA for fiscal year 2017. We described the IG reported maturity levels, including the Office of Inspectors General FISMA Reporting Metrics definition of "effectiveness." These maturity levels are based on security domains aligned with the five core functions in NIST's cybersecurity framework. We also summarized IG reported conclusions on the effectiveness of agencies' information security programs for fiscal year 2017.
- reviewed the fiscal year 2016 and 2017 financial statement audit reports for each of the 23 civilian CFO Act agencies to identify the extent to which any significant deficiencies or material weaknesses related to information security over financial systems had been reported and to identify information security control weaknesses reported by the IGs.
- identified whether agencies had met the targets for the cybersecurity focused cross-agency priority goal for fiscal years

---

[84] The *Federal Information Security Modernization Act of 2014* (FISMA 2014), enacted as Pub. L. No. 113-283, 128 Stat. 3073 (Dec. 18, 2014), largely superseded the *Federal Information Security Management Act of 2002* (FISMA 2002), enacted as *Title III, E-Government Act of 2002*, Pub. L. No. 107-347, 116 Stat. 2899, 2946 (Dec. 17, 2002). As used in this report, FISMA refers both to FISMA 2014 and to those provisions of FISMA 2002 that were either incorporated into FISMA 2014 or were unchanged and continue in full force and effect.

[85] The White House, *Strengthening the Cybersecurity of Federal Networks and Critical Infrastructure,* Executive Order 13800 (Washington, D.C.: May 11, 2017), 82 Fed. Reg. 22391 (May 16, 2017).

[86] National Institute of Standards and Technology, *Framework for Improving Critical Infrastructure Cybersecurity*, Version 1.1 (Gaithersburg, MD: Apr. 16, 2018).

2016 and 2017 by examining agency-reported performance metrics for fiscal years 2016 and 2017.

- evaluated OMB's agency risk management assessment ratings to make a determination on how agencies were managing risk to their enterprise. These conclusions were based on FISMA metrics, and are aligned with the five core security functions defined in the cybersecurity framework.
- interviewed knowledgeable OMB officials and staff to obtain their views on the reported effectiveness of the federal government's approach and strategy to securing agency information systems.

To assess the extent to which OMB and DHS have facilitated the use of intrusion detection and prevention capabilities to secure federal agency information systems, we

- determined the extent OMB and DHS fulfilled their requirements described in the *Federal Cybersecurity Enhancement Act of 2015* by collecting and reviewing artifacts from OMB and DHS and comparing them to the provisions outlined in the act. We also interviewed knowledgeable officials from OMB and DHS regarding their efforts to fulfill their requirements described in the act.
- determined the effectiveness of corrective actions taken by DHS to address nine previously reported recommendations we made in our report related to NCPS.[87] Specifically, we collected appropriate artifacts and assessed the artifacts against the criteria used in that report, and determined the extent to which the actions taken by DHS met the intent of the recommendations, and we met with DHS staff responsible for the remediation activities and obtained their views of the status of actions taken to address the recommendations.

---

[87] GAO, *Information Security: DHS Needs to Enhance Capabilities, Improve Planning, and Support Greater Adoption of Its National Cybersecurity Protection System,* GAO-16-294 (Washington, D.C.: Jan. 28, 2016).

- held semi-structured interviews[88] with knowledgeable officials from the 23 civilian CFO Act agencies. During these interviews, we obtained the agency's views on whether they need more training and guidance from DHS for NCPS and CDM. We also interviewed knowledgeable officials and staff at DHS to obtain their views on how DHS had improved the intrusion detection and prevention capabilities it provides to federal agencies. We also interviewed DHS officials to obtain their views on the training and guidance that the department makes available to agencies.

To assess the extent to which selected agencies reported implementing intrusion detection and prevention capabilities, we described the reported intrusion detection and prevention capabilities implemented by the 23 civilian CFO Act civilian agencies by

- summarizing implemented intrusion detection and prevention capability information obtained from the semi-structured interviews at the 23 civilian CFO Act agencies described above;
- identifying the extent to which the 23 civilian CFO Act agencies were in compliance with DHS's binding operating directive (BOD) pertaining to enhanced email and web security (BOD 18-01) by collecting and summarizing *Cyber Hygiene Trustworthy Email* reports from the 23 agencies and determining the extent to which the agencies had taken required actions to implement the BOD.

During the semi-structured interviews, we also obtained the agency's views and experiences with other programs and services provided by DHS, including the extent to which agencies had implemented the tools offered by the department's Continuous Diagnostics and Mitigation (CDM) program.

---

[88] A semi-structured interview methodology generally involves asking a similar subset of questions of multiple interviewees. We used a semi-structured interview format with both closed- and open-ended questions. The intent of our open-ended questions was to engage the agency officials in a conversation about the topics being discussed.

To determine the reliability of submitted data and obtain clarification about agencies' processes to ensure the accuracy and completeness of data used in their respective FISMA reports, we analyzed documents and conducted interviews with officials from 6 of the 23 civilian CFO Act agencies. To select these six agencies, we sorted agency fiscal year 2017 information technology budget data from highest to lowest amount and then divided the data into three tiers: high spending, medium spending, and low spending. We then randomly selected two agencies from each of the three tiers. The selected agencies were the Departments of Agriculture, Commerce, Housing and Urban Development, Transportation, and Veterans Affairs, and the U.S. Agency for International Development. While not generalizable to all agencies, the information we collected and analyzed about the six selected agencies provided insights into various processes in place to produce FISMA reports. Based on this assessment, we determined that the data were sufficiently reliable for the purposes of our reporting objectives.

We conducted this performance audit from December 2017 to December 2018 in accordance with generally accepted government auditing standards. Those standards require that we plan and perform the audit to obtain sufficient, appropriate evidence to provide a reasonable basis for our findings and conclusions based on our audit objectives. We believe that the evidence obtained provides a reasonable basis for our findings and conclusions based on our audit objectives.

APPENDIX II: CYBERSECURITY FRAMEWORK

The National Institute of Standards and Technology established the cybersecurity framework to provide guidance for cybersecurity activities within the private sector and government agencies at all levels.[89] The cybersecurity framework consists of five core functions: identify, protect, detect, respond, and recover. Within the five functions are 23 categories

[89] National Institute of Standards and Technology, *Framework for Improving Critical Infrastructure Cybersecurity*, Version 1.1 (Gaithersburg, MD: Apr. 16, 2018).

and 108 subcategories that define discrete outcomes for each function, as described in Table 7.

Table 7. National Institute of Standards and Technology Framework for Improving Critical Infrastructure Cybersecurity

Category	Subcategory
Identify (ID) core function	
Asset Management (ID.AM): The data, personnel, devices, systems, and facilities that enable the organization to achieve business purposes are identified and managed consistent with their relative importance to organizational objectives and the organization's risk strategy.	ID.AM-1: Physical devices and systems within the organization are inventoried.
	ID.AM-2: Software platforms and applications within the organization are inventoried.
	ID.AM-3: Organizational communication and data flows are mapped.
	ID.AM-4: External information systems are catalogued.
	ID.AM-5: Resources (e.g., hardware, devices, data, time, personnel, and software) are prioritized based on their classification, criticality, and business value.
	ID.AM-6: Cybersecurity roles and responsibilities for the entire workforce and third-party stakeholders (e.g., suppliers, customers, partners) are established.
Business Environment (ID.BE): The organization's mission, objectives, stakeholders, and activities are understood and prioritized; this information is used to inform cybersecurity roles, responsibilities, and risk management decisions	ID.BE-1: The organization's role in the supply chain is identified and communicated.
	ID.BE-2: The organization's place in critical infrastructure and its industry sector is identified and communicated.
	ID.BE-3: Priorities for organizational mission, objectives, and activities are established and communicated.
	ID.BE-4: Dependencies and critical functions for delivery of critical services are established.
	ID.BE-5: Resilience requirements to support delivery of critical services are established for all operating states (e.g., under duress/attack, during recovery, normal operations).
Governance (ID.GV): The policies, procedures, and processes to manage and monitor the organization's regulatory, legal, risk, environmental, and operational requirements are understood and inform the management of cybersecurity risk.	ID.GV-1: Organizational cybersecurity policy is established and communicated.
	ID.GV-2: Cybersecurity roles and responsibilities are coordinated and aligned with internal roles and external partners.
	ID.GV-3: Legal and regulatory requirements regarding cybersecurity, including privacy and civil liberties obligations, are understood and managed.
	ID.GV-4: Governance and risk management processes address cybersecurity risks.

Table 7. (Continued)

Category	Subcategory
Risk Assessment (ID.RA): The organization understands the cybersecurity risk to organizational operations (including mission, functions, image, or reputation), organizational assets, and individuals.	ID.RA-1: Asset vulnerabilities are identified and documented.
	ID.RA-2: Cyber threat intelligence is received from information sharing forums and sources.
	ID.RA-3: Threats, both internal and external, are identified and documented.
	ID.RA-4: Potential business impacts and likelihoods are identified.
	ID.RA-5: Threats, vulnerabilities, likelihoods, and impacts are used to determine risk.
	ID.RA-6: Risk responses are identified and prioritized.
Risk Management Strategy (ID.RM): The organization's priorities, constraints, risk tolerances, and assumptions are established and used to support operational risk decisions.	ID.RM-1: Risk management processes are established, managed, and agreed to by organizational stakeholders.
	ID.RM-2: Organizational risk tolerance is determined and clearly expressed.
	ID.RM-3: The organization's determination of risk tolerance is informed by its role in critical infrastructure and sector specific risk analysis.
Supply Chain Risk Management (ID.SC): The organization's priorities, constraints, risk tolerances, and assumptions are established and used to support risk decisions associated with managing supply chain risk. The organization has established and implemented the processes to identify, assess and manage supply chain risks.	ID.SC-1: Cyber supply chain risk management processes are identified, established, assessed, managed, and agreed to by organizational stakeholders.
	ID.SC-2: Suppliers and third party partners of information systems, components, and services are identified, prioritized, and assessed using a cyber supply chain risk assessment process.
	ID.SC-3: Contracts with suppliers and third-party partners are used to implement appropriate measures designed to meet the objectives of an organization's cybersecurity program and Cyber Supply Chain Risk Management Plan.
	ID.SC-4: Suppliers and third-party partners are routinely assessed using audits, test results, or other forms of evaluations to confirm they are meeting their contractual obligations.
	ID.SC-5: Response and recovery planning and testing are conducted with suppliers and third-party providers.
Identity Management, Authentication and Access Control (PR.AC): Access to physical and logical assets and associated facilities is limited to authorized users, processes, and devices, and is managed consistent with the assessed risk of unauthorized access to authorized activities and	PR.AC-1: Identities and credentials are issued, managed, verified, revoked, and audited for authorized devices, users and processes.
	PR.AC-2: Physical access to assets is managed and protected.
	PR.AC-3: Remote access is managed.
	PR.AC-4: Access permissions and authorizations are managed, incorporating the principles of least privilege and separation of duties.

Category	Subcategory
transactions.	
Protect (PR) core function	
	PR.AC-5: Network integrity is protected (e.g., network segregation, network segmentation).
	PR.AC-6: Identities are proofed and bound to credentials and asserted in interactions.
	PR.AC-7: Users, devices, and other assets are authenticated (e.g., single-factor, multi-factor) commensurate with the risk of the transaction (e.g., individuals' security and privacy risks and other organizational risks)
Awareness and Training (PR.AT): The organization's personnel and partners are provided cybersecurity awareness education and are trained to perform their cybersecurity-related duties and responsibilities consistent with related policies, procedures, and agreements.	PR.AT-1: All users are informed and trained.
	PR.AT-2: Privileged users understand their roles and responsibilities.
	PR.AT-3: Third-party stakeholders (e.g., suppliers, customers, partners) understand their roles and responsibilities.
	PR.AT-4: Senior executives understand their roles and responsibilities.
	PR.AT-5: Physical and cybersecurity personnel understand their roles and responsibilities.
Data Security (PR.DS): Information and records (data) are managed consistent with the organization's risk strategy to protect the confidentiality, integrity, and availability of information.	PR.DS-1: Data-at-rest is protected.
	PR.DS-2: Data-in-transit is protected.
	PR.DS-3: Assets are formally managed throughout removal, transfers, and disposition.
	PR.DS-4: Adequate capacity to ensure availability is maintained.
	PR.DS-5: Protections against data leaks are implemented.
	PR.DS-6: Integrity checking mechanisms are used to verify software, firmware, and information integrity.
	PR.DS-7: The development and testing environment(s) are separate from the production environment.
	PR.DS-8: Integrity checking mechanisms are used to verify hardware integrity.
Information Protection Processes and Procedures (PR.IP): Security policies (that address purpose, scope, roles, responsibilities, management commitment, and coordination among organizational entities), processes, and procedures are maintained and used to manage protection of information systems and assets.	PR.IP-1: A baseline configuration of information technology/industrial control systems is created and maintained incorporating security principles (e.g., concept of least functionality).
	PR.IP-2: A System Development Life Cycle to manage systems is implemented.
	PR.IP-3: Configuration change control processes are in place.
	PR.IP-4: Backups of information are conducted, maintained, and tested.
	PR.IP-5: Policy and regulations regarding the physical operating environment for organizational assets are met.

Table 7. (Continued)

Category	Subcategory
	PR.IP-6: Data is destroyed according to policy.
	PR.IP-7: Protection processes are improved.
	PR.IP-8: Effectiveness of protection technologies is shared.
	PR.IP-9: Response plans (Incident Response and
	Business Continuity) and recovery plans (Incident Recovery and Disaster Recovery) are in place and managed.
	PR.IP-10: Response and recovery plans are tested.
	PR.IP-11: Cybersecurity is included in human resources practices (e.g., deprovisioning, personnel screening).
	PR.IP-12: A vulnerability management plan is developed and implemented.
Maintenance (PR.MA): Maintenance and repairs of industrial control and information system components are performed consistent with policies and procedures.	PR.MA-1: Maintenance and repair of organizational assets are performed and logged, with approved and controlled tools.
	PR.MA-2: Remote maintenance of organizational assets is approved, logged, and performed in a manner that prevents unauthorized access.
Protective Technology (PR.PT): Technical security solutions are managed to ensure the security and resilience of systems and assets, consistent with related policies, procedures, and agreements.	PR.PT-1: Audit/log records are determined, documented, implemented, and reviewed in accordance with policy.
	PR.PT-2: Removable media is protected and its use restricted according to policy.
	PR.PT-3: The principle of least functionality is incorporated by configuring systems to provide only essential capabilities.
	PR.PT-4: Communications and control networks are protected.
	PR.PT-5: Mechanisms (e.g., failsafe, load balancing, hot swap) are implemented to achieve resilience requirements in normal and adverse situations.
Detect (DE) core function	
Anomalies and Events (DE.AE): Anomalous activity is detected and the potential impact of events is understood.	DE.AE-1: A baseline of network operations and expected data flows for users and systems is established and managed.
	DE.AE-2: Detected events are analyzed to understand attack targets and methods.
	DE.AE-3: Event data are collected and correlated from multiple sources and sensors.
	DE.AE-4: Impact of events is determined.
	DE.AE-5: Incident alert thresholds are established.
Security Continuous Monitoring (DE.CM): The information	DE.CM-1: The network is monitored to detect potential cybersecurity events.

Category	Subcategory
system and assets are monitored to identify cybersecurity events and verify the effectiveness of protective measures.	DE.CM-2: The physical environment is monitored to detect potential cybersecurity events.
	DE.CM-3: Personnel activity is monitored to detect potential cybersecurity events.
	DE.CM-4: Malicious code is detected.
	DE.CM-5: Unauthorized mobile code is detected.
	DE.CM-6: External service provider activity is monitored to detect potential cybersecurity events.
	DE.CM-7: Monitoring for unauthorized personnel, connections, devices, and software is performed.
	DE.CM-8: Vulnerability scans are performed.
Detection Processes (DE.DP): Detection processes and procedures are maintained and tested to ensure awareness of anomalous events.	DE.DP-1: Roles and responsibilities for detection are well defined to ensure accountability.
	DE.DP-2: Detection activities comply with all applicable requirements.
	DE.DP-3: Detection processes are tested.
	DE.DP-4: Event detection information is communicated.
	DE.DP-5: Detection processes are continuously improved.
Respond (RS) core function	
Response Planning (RS.RP): Response processes and procedures are executed and maintained, to ensure response to detected cybersecurity incidents.	RS.RP-1: Response plan is executed during or after an incident.
Communications (RS.CO): Response activities are coordinated with internal and external stakeholders (e.g., external support from law enforcement agencies).	RS.CO-1: Personnel know their roles and order of operations when a response is needed.
	RS.CO-2: Incidents are reported consistent with established criteria.
	RS.CO-3: Information is shared consistent with response plans.
	RS.CO-4: Coordination with stakeholders occurs consistent with response plans.
	RS.CO-5: Voluntary information sharing occurs with external stakeholders to achieve broader cybersecurity situational awareness.
Respond (RS) core function	
Analysis (RS.AN): Analysis is conducted to ensure effective response and support recovery activities.	RS.AN-1: Notifications from detection systems are investigated.
	RS.AN-2: The impact of the incident is understood.
	RS.AN-3: Forensics are performed.
	RS.AN-4: Incidents are categorized consistent with response plans.
	RS-AN-5: Processes are established to receive, analyze and respond to vulnerabilities disclosed to the organization from internal and external sources (e.g., internal testing, security bulletins, or security researchers).

Table 7. (Continued)

Category	Subcategory
Mitigation (RS.MI): Activities are performed to prevent expansion of an event, mitigate its effects, and resolve the incident.	RS.MI-1: Incidents are contained.
	RS.MI-2: Incidents are mitigated.
	RS.MI-3: Newly identified vulnerabilities are mitigated or documented as accepted risks.
Improvements (RS.IM): Organizational response activities are improved by incorporating lessons learned from current and previous detection/response activities.	RS.IM-1: Response plans incorporate lessons learned.
	RS.IM-2: Response strategies are updated
Recovery Planning (RC.RP): Recovery processes and procedures are executed and maintained to ensure restoration of systems or assets affected by cybersecurity incidents.	RC.RP-1: Recovery plan is executed during or after a cybersecurity incident.
Improvements (RC.IM): Recovery planning and processes are improved by incorporating lessons learned into future activities.	RC.IM-1: Recovery plans incorporate lessons learned.
	RC.IM-2: Recovery strategies are updated.
Communications (RC.CO): Restoration activities are coordinated with internal and external parties (e.g., coordinating centers, Internet Service Providers, owners of attacking systems, victims, other CSIRTs, and vendors).	RC.CO-1: Public relations are managed.
	RC.CO-2: Reputation is repaired after an incident.
	RC.CO-3: Recovery activities are communicated to internal and external stakeholders as well as executive and management teams.

Source: National Institute of Standards and Technology. | GAO-19-105.

APPENDIX III: REPORTED EFFECTIVENESS OF AGENCIES' IMPLEMENTATION OF THE FEDERAL APPROACH FOR SECURING INFORMATION SYSTEMS

The federal approach and strategy for securing information systems is prescribed by federal law and policy, including the *Federal Information*

Security Modernization Act of 2014[90] and the presidential executive order on *Strengthening the Cybersecurity of Federal Networks and Critical Infrastructure*.[91] Accordingly, federal reports describing agency implementation of this law and policy, and reports of related agency information security activities, indicated the effectiveness of agencies' efforts to implement the federal approach and strategy. Table 8 summarizes the reported effectiveness of the 23 civilian *Chief Financial Officers Act of 1990* agencies to implement the government's approach and strategy to securing information systems.

Table 8. Fiscal Year 2017 Indicators of the 23 Selected Civilian *Chief Financial Officers Act of 1990* Agencies' Effectiveness in Implementing the Federal Approach and Strategy for Securing Information Systems

Agency	Inspector General Information Security Program Ratings	Inspector General Internal Control Deficiencies over Financial Reporting	CIO Cybersecurity Cross-Agency Priority Goal Targets	OMB Risk Management Assessment Ratings
US Department of Agriculture	Not effective	Material weakness	Not met	At risk
Department of Commerce	Not effective	Significant deficiency	Not met	At risk
Department of Education	Not effective	Significant deficiency	Met	Managing risk
Department of Energy	Effective	—	Not met	At risk
Department of Health and Human Services	Not effective	Material weakness	Not met	At risk

[90] The *Federal Information Security Modernization Act of 2014*, enacted as Pub. L. No. 113-283, 128 Stat. 3073 (Dec. 18, 2014), largely superseded the *Federal Information Security Management Act of 2002*, enacted as *Title III, E-Government Act of 2002*, Pub. L. No. 107-347, 116 Stat. 2899, 2946 (Dec. 17, 2002).

[91] The White House, *Strengthening the Cybersecurity of Federal Networks and Critical Infrastructure*, Executive Order 13800 (Washington, D.C.: May 11, 2017), 82 Fed. Reg. 22391 (May 16, 2017).

Table 8. (Continued)

Agency	Inspector General Information Security Program Ratings	Inspector General Internal Control Deficiencies over Financial Reporting	CIO Cybersecurity Cross-Agency Priority Goal Targets	OMB Risk Management Assessment Ratings
Department of Homeland Security	Effective	Material weakness	Not met	Managing risk
Department of Housing and Urban Development	Not effective	Significant deficiency	Met	Managing risk
Department of the Interior	Not effective	Significant deficiency	Not met	Managing risk
Department of Justice	Not effective	—	Not met	Managing risk
Department of Labor	Not effective	Significant deficiency	Not met	Managing risk
Department of State	Not effective	Significant deficiency	Not met	At risk
Department of Transportation	Not effective	—	Not met	At risk
Department of the Treasury	Not effective	Material weakness	Not met	Managing risk
Department of Veterans Affairs	Not effective	Material weakness	Not met	At risk
Environmental Protection Agency	Effective	Significant deficiency	Not met	At risk
General Services Administration	Not effective	Significant deficiency	Met	Managing risk
National Aeronautics and Space Administration	Not effective	Significant deficiency	Not met	At risk
National Science Foundation	Effective	—	Not met	Managing risk
Nuclear Regulatory Commission	Effective	—	Not met	Managing risk
Office of Personnel Management	Not effective	Material weakness	Met	Managing risk

Agency	Inspector General Information Security Program Ratings	Inspector General Internal Control Deficiencies over Financial Reporting	CIO Cybersecurity Cross-Agency Priority Goal Targets	OMB Risk Management Assessment Ratings
Small Business Administration	Not effective	Significant deficiency	Not met	At risk
Social Security Administration	Not effective	Significant deficiency	Met	Managing risk
US Agency for International Development	Effective	—	Met	Managing risk
Totals[a]	17[b]	17c	17d	10e

Legend: "—" means that information security was not designated as a significant deficiency or material weakness for that agency.

Source: GAO analysis based on agency and Office of Management and Budget fiscal year 2017 *Federal Information Security Modernization Act* data and agency financial reports for fiscal year 2017. | GAO-19-105.

[a]Although the totals for some of the columns are the same, the agencies included in the totals are not.

[b]The inspector general for 17 agencies reported that their agency did not have an effective information security program.

[c]The inspector general for 17 agencies designated information security as a significant deficiency or material weakness.

[d]The chief information officers for 17 agencies reported that their agencies did not meet all nine targets for the cybersecurity cross-agency priority goal.

[e]OMB reported that 10 agencies had enterprises that were at risk.

APPENDIX IV: UPDATED CYBERSECURITY-FOCUSED CROSS-AGENCY PRIORITY GOAL

The President's Management Agenda identifies cross-agency priority (CAP) goals to target areas where multiple agencies must collaborate to effect change.[92]

[92] The President's Management Agenda is intended to lay out a long-term vision for modernizing the federal government in key areas that will improve the ability of agencies to deliver mission outcomes, provide excellent service, and effectively steward taxpayer dollars on behalf of the American people.

The agenda issued in fiscal year 2018 established an information technology modernization goal that includes a cybersecurity objective with specific priority areas and performance indicators.

This cybersecurity-focused goal is intended to drive progress in the government's efforts to modernize information technology to increase productivity and security. Figure 8 describes the 3 updated cybersecurity-focused cross-agency priority areas and 10 performance indicators. Each federal agency is expected to meet one of the 10 new performance indicators by the end of fiscal year 2018 and the remainder by 2020.

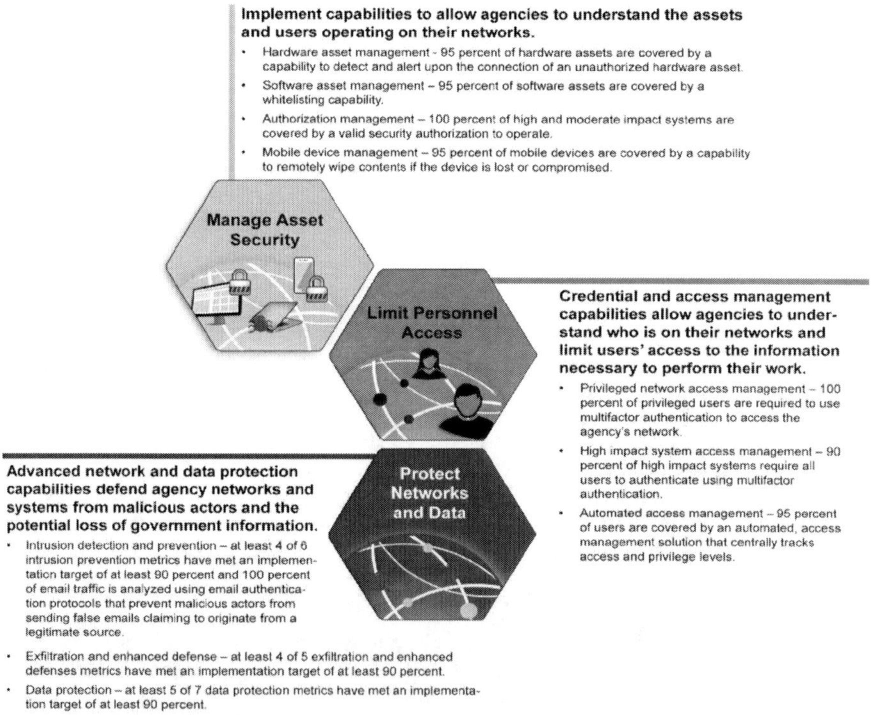

Source: Performance.gov. | GAO-19-105.

Figure 8. Cybersecurity-Focused Cross-Agency Priority Goal Priority Areas and Performance Indicators, Fiscal Years 2018– 2022.

Appendix V: Comments from the Department of Homeland Security

U.S. Department of Homeland Security
Washington, DC 20528

Homeland Security

December 11, 2018

Gregory C. Wilshusen
Director, Information Security Issues
U.S. Government Accountability Office
441 G Street, NW
Washington, DC 20548

Re: Management Response to Draft Report GAO-19-105, "INFORMATION SECURITY: Agencies Need to Improve Implementation of Federal Approach to Securing Systems and Protecting Against Intrusions" (Job Code 102490)

Dear Mr. Wilshusen,

Thank you for the opportunity to review and comment on this draft report. The U.S. Department of Homeland Security (DHS) appreciates the U.S. Government Accountability Office's (GAO) work in planning and conducting its review and issuing this report.

The Department is pleased to note GAO's positive recognition of DHS initiatives being taken to further secure agency information systems. These include the development of an in-depth "Intrusion Assessment Plan" that outlines tools, platforms, and resources; development of the National Cybersecurity Protection System (NCPS); and providing tools and services to agencies to monitor their networks through the Continuous Diagnostic and Mitigation (CDM) program. DHS is committed to helping agencies, which are responsible for their own cybersecurity, reduce the risk of successful cyber-attacks.

The draft report contained nine recommendations including two for DHS with which the Department concurs. Attached find our detailed response to each recommendation. Technical comments were previously provided under separate cover.

Again, thank you for the opportunity to review and comment on this draft report. Please feel free to contact me if you have any questions. We look forward to working with you again in the future.

Sincerely,

JIM H. CRUMPACKER, CIA, CFE
Director
Departmental GAO-OIG Liaison Office

Attachment: Management Response to the Recommendations Contained in GAO-19-105

GAO recommended that the Secretary of Homeland Security:

Recommendation 1: Direct the Network Security Deployment division to coordinate further with federal agencies to identify training and guidance needs for implementing NCPS and CDM.

Response: Concur. The Cybersecurity and Infrastructure Security Agency's (CISA) NCPS Program Office will work with the CISA's Federal Network Resilience (FNR) Division to ensure a broader dissemination of existing training materials and offerings to all federal agencies. In addition, the NCPS Program Office will explore options for improving the training offerings that are available. Estimated Completion Date (ECD): November 30, 2019.

Recommendation 2: Direct the appropriate staff to work with Office of Management and Budget (OMB) to follow-up with agencies to identify obstacles, and impediments affecting their abilities to implement intrusion detection and prevention capabilities.

Response: Concur. The CISA NCPS Program Office and FNR Division will continue to work with OMB and federal agencies to identify obstacles and impediments to effectively implementing NCPS capabilities. The NCPS Program Office meets with representatives from OMB on a bi-weekly basis to discuss the status of agency efforts. As part of these bi-weekly meetings, CISA staff will work with OMB to develop an approach to better engage on follow-up activities. ECD: September 30, 2019.

APPENDIX VI: COMMENTS FROM THE DEPARTMENT OF COMMERCE

UNITED STATES DEPARTMENT OF COMMERCE
Chief Information Officer
Washington, D.C. 20230

DEC 1 0 2018

Mr. Gregory C. Wilshusen
Director, Information Security Issues
Government Accountability Office
441 G Street NW
Washington, DC 20548

Dear Mr. Wilshusen:

Thank you for the opportunity to review and comment on the Government Accountability Office's (GAO) draft report entitled *Information Security: Agencies Need to Improve Implementation of Federal Approach to Securing Systems and Protecting Against Intrusions* (GAO-19-105).

On behalf of the Department of Commerce, I have enclosed our comments on the report. We concur and have no comments on the draft report.

Sincerely,

RODNEY TURK Digitally signed by RODNEY TURK Date: 2018.12.10 09:50:11 -05'00'

Rodney W. Turk
Acting Chief Information Officer

Department of Commerce
Office of the Chief Information Officer
Office of the Secretary

Technical and Editorial Comments on the GAO Report Entitled Information Security: Agencies Need to Improve Implementation of Federal Approach to Securing Systems and Protecting Against Intrusions (GAO-19-105).

The Office of the Chief Information Officer has reviewed the report and provides general comments are below. The OCIO has no technical and editorial comments, also reflected below. Page numbers refer to page numbers in the report unless otherwise stated.

General Comments
The report is reasonable, and we concur with its findings and recommendations.

Recommended Changes for Factual/Technical Information.

None

Editorial Comments

None

APPENDIX VII: COMMENTS FROM THE SOCIAL SECURITY ADMINISTRATION

SOCIAL SECURITY
Office of the Commissioner

December 10, 2018

Mr. Gregory C. Wilshusen
Director, Information Security Issues
United States Government Accountability Office
441 G Street, NW
Washington, DC 20548

Dear Mr. Wilshusen:

Thank you for the opportunity to review the draft report, "INFORMATION SECURITY: Agencies Need to Improve Implementation of Federal Approach to Securing Systems and Protecting Against Intrusions" (GAO-19-105). Please see our enclosed comments.

Sincerely,

Stephanie Hall

Stephanie Hall
Acting Deputy Chief of Staff

SSA COMMENTS ON THE GOVERNMENT ACCOUNTABILITY OFFICE (GAO) DRAFT REPORT, "INFORMATION SECURITY: AGENCIES NEED TO IMPROVE IMPLEMENTATION OF FEDERAL APPROACH TO SECURING SYSTEMS AND PROTECTING AGAINST INTRUSIONS" (GAO-19-105)

Protecting our networks and the information we use to administer our programs is a critical priority. We can never rest in our efforts to keep the public's data secure. We address this challenge proactively with an integrated, multi-layered, risk-based program that continually adds new defenses. We work with the Department of Homeland Security and third-party assessors to test the effectiveness of our program and identify areas of improvement. We are making ongoing improvements to our information security protocols to keep pace with changes in the operating environment and to mitigate known risks.

We continually improve our processes and capabilities to address the ever-changing threat environment and escalating risks. In fiscal year 2018, we made improvements and progress in securing applications, leveraging the cloud, managing our assets and vulnerabilities, strengthening our network and incident response capabilities, improving our security training, and enhancing the overall effectiveness of our cybersecurity program. We are confident that our proactive planning, coupled with our responsiveness to external assessments and continuous improvement processes, should provide the foundation for achieving a higher Federal Information Security Management Act compliance maturity level.

APPENDIX VIII: COMMENTS FROM THE U.S. AGENCY FOR INTERNATIONAL DEVELOPMENT

USAID
FROM THE AMERICAN PEOPLE

Gregory C. Wilshusen
Director, Information Security Issues
United States Government Accountability Office
441 G Street, N.W.
Washington, D.C. 20548

Re: INFORMATION SECURITY: Agencies Need to Improve Implementation of Federal Approach to Securing Systems and Protecting Against Intrusions (GAO-19-105)

Dear Mr. Wilshusen:

I am pleased to provide the formal response of the U. S. Agency for International Development (USAID) to the draft report produced by the U. S. Government Accountability Office (GAO) entitled, *"INFORMATION SECURITY: Agencies Need to Improve Implementation of Federal Approach to Securing Systems and Protecting Against Intrusions" (GAO-19-105).*

As the report indicates, USAID has a successful approach to securing our information-technology (IT) systems and protecting against intrusion. Table 8 in the report (pages 62-63) lists USAID as the only Department or Agency within the Federal Government rated as effectively meeting the Inspector General Information-Security Program Ratings, targets of the Chief Information Officer's Cybersecurity Cross-Agency Priority Goal, and the Office of Management and Budget's Risk-Management Assessment Ratings.

The Agency continues to mature its cybersecurity posture, as shown by the improvements seen in the USAID 2018 Office of Inspector General Annual Federal Information Security Management Act (FISMA) Metrics Report, based on 2018 data. Since then, USAID has improved in three Core Security Functions of the Inspector General Ratings of Agencies' Information-Security Policies, Procedures, and Practices (pg. 22):

- For Fiscal Year (FY) 2018, USAID was rated at a 5 for "Identify";
- For FY 2018, USAID was rated at a 4 for "Detect"; and
- For FY 2018, USAID was rated at a 4 for "Recover."

I am transmitting this letter for inclusion in the GAO's final report. Thank you for the opportunity to respond to the draft report, and for the courtesies extended by your staff while conducting this engagement.

Sincerely,

Angelique M. Crumbly
Acting Assistant Administrator
Bureau for Management

In: Information Security
Editor: Paul Scholz

ISBN: 978-1-53615-880-9

Chapter 7

DATA BREACHES: RANGE OF CONSUMER RISKS HIGHLIGHTS LIMITATIONS OF IDENTITY THEFT SERVICES*

United States Government Accountability Office

ABBREVIATIONS

CFPB	Consumer Financial Protection Bureau
FTC	Federal Trade Commission
IRS	Internal Revenue Service
OMB	Office of Management and Budget
OPM	Office of Personnel Management
PIN	personal identification number

* This is an edited, reformatted and augmented version of the United States Government Accountability Office Report to Congressional Requesters, Publication No. GAO-19-230, dated March 2019.

Why GAO Did This Study

Recent large-scale data breaches of public and private entities have put hundreds of millions of people at risk of identity theft or other harm. GAO was asked to review issues related to consumers' options to address risks of harm from data breaches. This chapter, among other things, examines information and expert views on the effectiveness of consumer options to address data breach risks. GAO analyzed available data on options, collected and analyzed related documentation, conducted a literature review of studies, and interviewed a nongeneralizable sample of 35 experts (from academia, government entities, consumer and industry organizations) and identity theft service providers to reflect a range of views.

What GAO Recommends

GAO reiterates a matter for congressional consideration and a recommendation from its 2017 report on identity theft services (GAO-17-254). In that report, GAO found that legislation requiring federal agencies that experience data breaches, including OPM, to offer certain levels of identity theft insurance coverage to affected individuals requires coverage levels that are likely unnecessary. Therefore, Congress should consider permitting agencies to determine the appropriate coverage level for such insurance. GAO also recommended the Office of Management and Budget (OMB) update its guidance for agency responses to data breaches, after analyzing the effectiveness of identity theft services relative to lower-cost alternatives. OMB did not agree or disagree and had not taken action as of early March 2019.

What GAO Found

No one solution can address the range of potential risks from a data breach, according to interviews with academic, consumer, government, and industry experts and documentation GAO reviewed. Perpetrators of fraud can use stolen personal information—such as account numbers, passwords, or Social Security numbers—to take out loans or seek medical care under someone else's name, or make unauthorized purchases on credit cards, among other crimes. Foreign state-based actors can use personal information to support espionage or other nefarious uses.

Public and private entities that experience a breach sometimes provide complimentary commercial identity theft services to affected individuals to help monitor their credit accounts or restore their identities in cases of identity theft, among other features. Consumers also may purchase the services. As of November 30, 2018, the Office of Personnel Management (OPM) had obligated about $421 million for a suite of credit and identity monitoring, insurance, and identity restoration services to offer to the approximately 22 million individuals affected by its 2015 data breaches. As of September 30, 2018, about 3 million had used the services and approximately 61 individuals had received payouts from insurance claims, for an average of $1,800 per claim. OPM re-competed and awarded a contract to the previously contracted company in December 2018.

GAO's review did not identify any studies that analyzed whether consumers who sign up for or purchase identity theft services were less subject to identity theft or detected financial or other fraud more or less quickly than those who monitored their own accounts for free. A few experts said consumers could sign up for such services if offered for free. Credit monitoring may be convenient for consumers and personalized restoration services may help identity theft victims recover their identities, but such services do not prevent fraud from happening in the first place. The services also do not prevent or directly address risks of nonfinancial harm such as medical identity theft.

Consumer, government, and industry experts highlighted other free options, including a credit freeze, which prevents one type of fraud. A freeze restricts businesses from accessing a person's credit report—and can prevent the illicit opening of a new account or loan in the person's name. A provision of federal law that took effect in September 2018 made it free for consumers to place or lift credit freezes quickly at the three nationwide consumer reporting agencies (Equifax, Experian, and TransUnion). Consumers also can regularly monitor their accounts and review their credit reports for free every 12 months. In addition, they can take advantage of free federal assistance such as the guidance on the Federal Trade Commission's IdentityTheft.gov website.

Finally, large amounts of personal information are outside of consumers' control and bad actors can use stolen information for years after a breach. Therefore, experts noted that data security at entities that hold such information—and efforts to make stolen information less useful for identity thieves, through use of new identity verification technologies, for example—are important ways to mitigate risks of harm for consumers.

March 27, 2019

The Honorable Frank Pallone
Chairman
Committee on Energy and Commerce
House of Representatives

The Honorable Jan Schakowsky
Chair
Subcommittee on Consumer Protection and Commerce
Committee on Energy and Commerce
House of Representatives

The Honorable Diana DeGette
Chair
Subcommittee on Oversight and Investigations
Committee on Energy and Commerce
House of Representatives

Sensitive personal information—such as Social Security numbers or dates of birth—can be exposed in several ways, including on a very large scale through a data breach.[1] For example, major retail and hotel chains have suffered data breaches that exposed the identifying information, including financial account or Social Security numbers, of millions of people. Consumers whose information is exposed can be at risk of a range of harms, including identity theft or fraud. Consumers can try to prevent or mitigate these harms in a number of ways, such as monitoring their credit reports and credit card statements for suspicious activity, or placing a credit freeze that restricts access to their credit report. They also may enroll in free or fee-based identity theft services.[2] Private-sector and government entities that experienced data breaches have provided these services to millions of affected consumers.

In March 2017, we reported on the potential benefits and limitations of commercially available identity theft services and factors that affect public- and private-sector decision-making about them.[3] Since that time, additional large-scale data breaches have occurred and Congress passed legislation that enhances some of the options available to consumers to prevent or mitigate identity theft. You asked us to review issues related to actions consumers can take to address risks of harm from data breaches. This chapter examines (1) information and expert views on the effectiveness of options consumers can use to prevent or address the risks resulting from data breaches; and (2) federal assistance available to help consumers understand these options, including the status of one matter for congressional consideration and one recommendation relating to these issues in our 2017 report.

---

[1] A data breach generally refers to an unauthorized or unintentional exposure, disclosure, or loss of an organization's sensitive information. This information can include personally identifiable information, such as Social Security numbers, or financial information, such as credit card numbers. A data breach can be inadvertent, such as from the loss of an electronic device, or deliberate, such as from the theft of a device or a cyber-based attack by a malicious individual or group, agency insiders, foreign nations, or terrorists.

[2] We use "identity theft services" to refer to commercial products that generally provide tools intended to help consumers detect identity theft and restore their identity if it has been compromised.

[3] GAO, *Identity Theft Services: Services Offer Some Benefits but Are Limited in Preventing Fraud*, GAO-17-254 (Washington, D.C.: Mar. 30, 2017).

To address the first objective, we conducted a literature review to identify any studies or independent research on the effectiveness of consumers' options for mitigating or preventing harm from exposure of personal information. We also searched for studies that examined consumer attitudes and behavior following data breaches, and harms to individuals from data breaches. We interviewed a nongeneralizable sample of experts and private companies that provide identity theft services to consumers.[4] Specifically, we interviewed representatives of 35 entities in the following categories: academic or independent research institution (4); consumer or privacy research and advocacy (10); industry association or identity theft service provider, or industry consultant (12); and federal or state government (9). We selected the experts and identity theft service providers to represent a range of perspectives. We also reviewed provisions in the Economic Growth, Regulatory Relief, and Consumer Protection Act, enacted in May 2018, that address credit freezes and fraud alerts (two tools for preventing one type of identity theft).[5] Furthermore, we reviewed the evidence collected for the 2017 GAO report on identity theft services.[6]

To address the second objective, we reviewed documentation and interviewed staff from the Federal Trade Commission (FTC), Consumer Financial Protection Bureau (CFPB), and Office of Personnel Management (OPM). We analyzed data from the company that contracted to provide identity theft services to individuals affected by two data breaches at OPM in 2015. We assessed the reliability of the data by interviewing agency officials and reviewing documentation about the systems used to store the data. We found the data to be reliable for

[4] Throughout this report, we use certain qualifiers when describing responses from interview participants, such as "few," "some," and "most." While we define few as a small number such as two or three, the specific quantification of other categories depends on the overall numbers of interviewees who addressed a specific topic, and is discussed in more detail in appendix I. We defined experts as those with academic research backgrounds or professional expertise gained from employment in consumer and industry policy organizations, as well as federal and state government staff with specific positions of responsibility in consumer protection.

[5] Pub. L. No. 115-174, § 301(a), 132 Stat. 1296, 1326 (2018) (codified at 15 U.S.C. § 1681c–1(i)).

[6] GAO-17-254.

purposes of this reporting objective. We also reviewed documentation and interviewed agency staff about the development, implementation, and assessment of their consumer education materials and other resources and assistance. We compared these activities against a 2014 Executive Order on the security of consumer financial transactions, key practices for consumer education planning identified in our prior work, and federal standards for internal control.[7] In addition, we followed up on recommendations made in our 2017 report. For more information on our scope and methodology, including the organization representatives we interviewed, see appendix I.

We conducted this performance audit from November 2017 to March 2019 in accordance with generally accepted government auditing standards. Those standards require that we plan and perform the audit to obtain sufficient, appropriate evidence to provide a reasonable basis for our findings and conclusions based on our audit objectives. We believe that the evidence obtained provides a reasonable basis for our findings and conclusions based on our audit objectives.

BACKGROUND

Harm from Exposure of Personal Information

Individuals' sensitive personal information can be lost, stolen, or given away.[8] Once exposed, individuals' information can be misused to commit

---

[7] Exec. Order No. 13681, 79 Fed. Reg. 63491 (Oct. 23, 2014). See GAO, Digital Television Transition: Increased Federal Planning and Risk Management Could Further Facilitate the DTV Transition, GAO-08-43 (Washington, D.C.: Nov. 19, 2007); and Standards for Internal Control in the Federal Government, GAO-14-704G (Washington, D.C.: Sept. 10, 2014). Also see Office of Management and Budget, Preparing for and Responding to a Breach of Personally Identifiable Information, M-17-12 (Washington, D.C.: Jan. 3, 2017).

[8] Identity thieves can obtain sensitive personal information through various methods. For instance, thieves can use phishing to trick individuals or employees of an organization into sharing their own or others' sensitive personal information. Phishing uses authentic-looking, but fake, emails to request information from users or direct them to a fake website that requests information. Identity theft also can occur as a result of the loss or theft of data (a lost or stolen wallet or a thief digging through household trash). Some individuals may

identity theft, fraud, or inflict other types of harm. Identity theft occurs when individuals' information is used without authorization in an attempt to commit fraud or other crimes. In 2016, according to the Bureau of Justice Statistics, an estimated 26 million people—10 percent of U.S. residents aged 16 or older—reported that they had been victims of identity theft in the previous year.[9] One potential source of identity theft is a data breach at an organization that maintains large amounts of sensitive personal information. Recent data breaches include the 2018 breach of Marriott International's Starwood guest registration database, which may have exposed information of millions of individuals, and the 2017 data breach at Equifax, Inc., a nationwide consumer reporting agency, which exposed identifying information of at least 145.5 million people.[10] The types of harm that can result from exposure of sensitive personal information include the following:

- Financial fraud from identity theft, which can include
    - *new-account fraud*, in which thieves use identifying data, such as Social Security and driver's license numbers, to open new financial accounts without that person's knowledge; and,
    - *existing-account fraud*, which is more common and entails the use or takeover of existing accounts, such as credit or debit card accounts, to make unauthorized charges or withdraw money.[11]

---

reveal sensitive information willingly, such as on social media accounts, which when combined with other information can allow fraudsters to steal identities.

[9] Department of Justice, Bureau of Justice Statistics, *Victims of Identity Theft, 2016* (Washington, D.C.: January 2019).

[10] We use "nationwide consumer reporting agency" to refer to the Fair Credit Reporting Act's term "consumer reporting agency that compiles and maintains files on consumers on a nationwide basis." The act defines such an agency as one that regularly engages in the practice of assembling or evaluating, and maintaining public record information and credit account information regarding consumers residing nationwide for the purpose of furnishing consumer reports to third parties bearing on a consumer's credit worthiness, credit standing, or credit capacity. 15 U.S.C. § 1681a(p). The three agencies that operate nationwide (Equifax, Experian, and TransUnion) provide reports commonly used to determine an individual's eligibility for credit, employment, and insurance.

[11] According to the Bureau of Justice Statistics, for 85 percent of identity-theft victims in 2016, the most recent incident involved misuse or attempted misuse of only one type of existing account, such as a credit card or bank account. Approximately 1 percent of those 16 or older experienced the opening of a new account or other misuse of personal information apart

- *Tax refund fraud*, which occurs when a Social Security number or other personally identifiable information is used to file a fraudulent tax return seeking a refund.[12]
- *Government benefits fraud*, which occurs when thieves use stolen personal information to fraudulently obtain government benefits. For example, the Social Security Administration has reported that personal information of beneficiaries has been used to fraudulently redirect the beneficiary's direct deposit benefits.[13]
- *Medical identity theft*, which occurs when someone uses an individual's name or personal identifying information to obtain medical services or prescription drugs fraudulently, including submitting fraudulent insurance claims.
- *Synthetic identity theft,* which involves the creation of a fictitious identity, typically by using a combination of real data and fabricated information. The federal government has identified synthetic identity theft as an emerging trend.[14]
- *Child identity theft*, which occurs when a child's Social Security number or other identifying information is stolen and used to commit fraudulent activity.

---

from misuse of an existing account. See Bureau of Justice Statistics, *Victims of Identity Theft, 2016.* While existing-account fraud is a significant problem, existing laws limit consumer liability for such fraud. As a matter of policy, some credit and debit card issuers may voluntarily cover all fraudulent charges. For example, for unauthorized credit card charges, cardholder liability is limited to a maximum of $50 per card. 15 U.S.C. § 1643; 12 C.F.R. § 1026.12. For unauthorized automated teller machine or debit card transactions, the Electronic Fund Transfer Act generally limits consumer liability, depending on how quickly the consumer reports the loss or theft of the card. *See* 15 U.S.C. § 1693g; 12 C.F.R. §1005.6.

[12] GAO, Identity Theft: IRS Needs to Strengthen Taxpayer Authentication Efforts, GAO-18-418 (Washington, D.C.: June 22, 2018).

[13] For a related GAO report, see Social Security Numbers: OMB Actions Needed to Strengthen Federal Efforts to Limit Identity Theft Risks by Reducing Collection, Use, and Display, GAO-17-553 (Washington, D.C.: July 25, 2017).

[14] GAO, Highlights of a Forum: Combating Synthetic Identity Fraud, GAO-17-708SP (Washington, D.C.: July 26, 2017). For example, individuals may face difficulties obtaining credit if their Social Security number has been used as part of a synthetic identity to commit fraud, or may face health risks if their records are connected to someone else. Synthetic identity fraud has grown significantly in recent years and resulted in significant financial losses to the financial industry and federal government.

- *Other types of fraud* that occur when personal information is used; for example, to set up mobile phone or utility accounts, or to engage in activities such as applying for employment or renting a home.

The harms caused by exposure of personal information or identity theft can extend beyond tangible financial loss, including the following:

- *Lost time.* Victims of identity theft or fraud may spend significant amounts of time working to restore their identities. In 2016, according to the Bureau of Justice Statistics survey of identity victims, most victims resolved issues in 1 day or less but about 1 percent of victims spent 6 months or more resolving their identity theft issues.[15]
- *Emotional distress and reputational harm.* Exposed information also can cause emotional distress, a loss of privacy, or reputational injury. In 2016, according to the Bureau of Justice Statistics, about 10 percent of those who experienced identity theft reported suffering severe emotional distress.
- *Harm from state-based actors.* State-sponsored espionage can cause harm to individuals when nations use cyber tools as part of information-gathering, espionage, or other nefarious activities.

Consumers' Options to Address Risks or Harm

Options available to consumers to help prevent or mitigate identity theft include actions they can take on their own (generally for free) or services they can purchase.[16]

Actions individual consumers can take themselves include the following:

---

[15] Department of Justice, Bureau of Justice Statistics, *Victims of Identity Theft, 2016* (January 2019).

[16] See GAO-17-254 for additional details on these options.

- *Placing a credit freeze.* A credit or security freeze restricts potential creditors from accessing a credit report until the consumer asks the agency to remove or temporarily lift the freeze.
- *Placing a fraud alert.* A fraud alert on a credit report requires businesses to verify a consumer's identity before they issue credit.
- *Monitoring accounts and other information.*[17]
    - *Reviewing free annual credit reports.* Individuals can request one copy of their credit report every 12 months (available for free at AnnualCreditReport.com).from each of the three nationwide consumer reporting agencies.[18]
    - *Reviewing financial statements and other accounts.* Individuals can review bank and other financial statements regularly for suspicious activity and make use of automatic transaction alerts and other free features that financial institutions offer to detect potential fraud. Individuals also can regularly review mobile phone or utility accounts for unusual activity.
    - *Reviewing health insurance benefits explanations and medical information.* Individuals can review explanation-ofbenefits statements from their health insurer to detect fraudulent insurance claims or monitor their files at their healthcare providers to detect unauthorized use of medical services.

Consumers also can obtain various free or fee-based *identity theft services*, which are commercial products that generally offer tools intended to help consumers detect identity theft and restore their identity if it has been compromised. The private research firm IBISWorld estimated that the U.S. market for identity theft services was about $3billion annually in

---

[17] Department of Justice, Bureau of Justice Statistics, *Victims of Identity Theft, 2016* (January 2019).The survey found that the most common ways victims discovered identity theft was by being contacted by a financial institution (about 48 percent) or noticing fraudulent charges on an account (about 19 percent). About 1.4 percent of victims said they discovered the theft through a credit report or credit monitoring service.

[18] Individuals can use other mechanisms to request and review credit reports from other consumer reporting agencies every year for free.

2015–2017.[19] The services may be marketed directly to individuals for a monthly or annual fee. In addition, private- and public-sector entities that have experienced data breaches sometimes purchase these services and offer them to affected individuals at no cost.[20]

- Identity theft services most often include
  - *credit monitoring*, which tracks an individual's credit reports and sends alerts about potentially suspicious activity;
  - *identity monitoring*, which aims to monitor other sources such as public records and illicit websites (sometimes referred to as the "dark web");
  - *identity restoration*, which provides a range of services to recover from identity theft; and
  - *identity theft insurance*, which reimburses individuals for certain costs related to the process of restoring identities.

Other actions consumers can take to protect their identity include adoption of certain data security practices and early filing of tax returns. Data security practices can help protect sensitive information. For example, individuals can change or avoid sharing or re-using passwords, and make use of strong passwords and authentication options on online accounts; properly safeguard or shred sensitive paper documents; and limit access to their sensitive information on social media. Filing a tax return early reduces the risk of tax refund fraud, and some victims of tax refund fraud may be eligible for an Identity Protection Personal Identification Number (PIN)—issued by the Internal Revenue Service IRS)—to prevent future fraud. To protect their Social Security benefits, individuals can set up an online account at the Social Security Administration to monitor their benefits accounts.

---

[19] See IBISWorld, *Identity Theft Protection Services in the US:* Industry Market Research Report. We cited the reports published in April 2015 and April 2016 in GAO-17-254; and accessed the August 2017 report at https://www.ibisworld.com/industry-trends/specialized-market-research-reports/technology/computer-services/identity-theft-protection-services.html, on August 30, 2018.

[20] For example, in response to breaches of its databases in 2015, OPM offered identity theft services to approximately 22 million people affected by the breaches.

Limited Information Is Available on Effectiveness of Options after Data Breaches, but Credit Freezes Can Prevent New-Account Fraud

We did not identify any studies that analyzed whether consumers who sign up for or purchase identity theft services encounter fewer instances of identity theft or detect instances of financial or other fraud more—or less—rapidly than consumers who take steps on their own. Views of experts varied, but most said identity theft services have limitations and would not address all data breach risks. Most experts also said that a credit freeze, which consumers place on their own for free, is a useful way to prevent one type of financial fraud—the illegal opening of new credit accounts in consumers' names. Based on our review and discussions with experts, consumers can consider four factors when deciding on options to address risks after a data breach: the extent to which an option might prevent fraud; the cost of an option; its convenience; and the type of information that was exposed and may be at risk.

No Independent Research Assesses Effectiveness of Consumer Options to Address Risks after Data Breaches

Information that can help consumers assess their options for mitigating and addressing the risks of identity theft and other harm from data breaches is limited. Specifically, we did not identify any studies that analyzed whether consumers who sign up for free or purchase identity theft services encounter fewer instances of identity theft or detect instances of financial or other fraud more—or less—rapidly than consumers who take steps on their own for free—such as monitoring their credit reports or placing a credit freeze. For consumers who experienced identity theft, we did not find any studies that compared the effectiveness of free options to help consumers recover from identity theft with commercial identity restoration services. In addition to searching databases of scholarly publications and other sources, a range of academic, consumer, government, and industry experts we interviewed told us that they were unaware of any specific independent studies on the effectiveness of consumer options.[21]

We interviewed representatives of seven companies that provide identity theft services about how they assess the effectiveness of their services and found that what they measure does not directly address how effective these services would be in mitigating the risks of identity theft compared with options consumers can take on their own. For example, two company representatives said that their services focus on detection of fraudulent activity or assistance after identity theft has occurred, rather than on prevention of identity theft or other harms. The representatives of each of the providers said that their companies generally measure how customers use their products and services; customer satisfaction (for example, through surveys or other feedback); and whether the products work as intended (for example, whether alerts of fraudulent activity are successfully delivered to customers or customers can successfully access the company's website when they need to). Companies that offer identity restoration services also measure the rate at which they complete the process of recovering stolen identities. While it is not possible to prevent identity fraud, four representatives said that early detection of fraud is important as it allows consumers to address potential fraud more quickly.

FTC, a primary source for assistance to consumers on issues related to data breaches and identity theft, has advised consumers that the effectiveness of services that offer identity monitoring depends on factors

---

[21] See Vyacheslav Mikhed and Michael Vogan, *How Data Breaches Affect Consumer Credit* (Philadelphia, Penn.: November 2017). The study used the 2012 South Carolina Department of Revenue data breach as a natural experiment to study how data breaches and news coverage about them affect consumers' interactions with the credit market and their use of credit. The study found that some consumers directly exposed to the breach protected themselves against potential losses from future fraudulent use of stolen information by monitoring their files and freezing access to their credit reports. The response of consumers only exposed to news about the breach was negligible. As part of their analysis, the researchers measured the extent to which affected individuals signed up for credit monitoring services, credit freezes, fraud alerts, or opted out of receiving prescreened offers of credit. The study found that the most frequent options chosen were credit monitoring services and credit freezes. The researchers also found that more individuals opted to place freezes on their credit reports than the researchers had predicted.

such as the kinds of databases the service provider monitors, how well the databases collect information, and how often the service provider checks each database.[22] For example, FTC suggests that consumers ask if service providers check databases that show payday loan applications or changes in addresses for misuse of their information as part of identity monitoring. In reviewing consumer education and promotional materials on the websites of five identity theft service companies we contacted that offer identity monitoring, we found that three providers included information about which types of databases they monitor; the other two did not.[23]

Government and commercial entities—such as federal agencies and retail stores—that decide to purchase identity theft services to offer to affected individuals after a breach of their data do not necessarily base their decision on how effective these services are. Rather, according to industry and some government representatives we interviewed, some base their decisions on federal or state legal requirements to offer such services and the expectations of affected customers or employees for some action on the breached entities' part. Representatives of retail and banking associations we interviewed indicated that it has become the industry standard to offer 1 year of credit or identity monitoring services in the wake of a data breach. One industry representative said that in some cases the decision is not based on the effectiveness of the services.[24] States such as California require companies to offer some type of identity theft service after a data breach.[25] Moreover, Connecticut requires health insurers and certain health care-related companies to offer identity theft services following an actual or suspected data breach. In 2017, we reported that companies do not assess

---

[22] Federal Trade Commission, *Consumer Information: Identity Theft Protection Services*, https://www.consumer.ftc.gov/articles/0235-identity-theft-protection-services; accessed on July 10, 2018.

[23] We reported in 2017 that the Consumer Federation of America offers guidance to individuals and entities on selecting identity theft service providers: Consumer Federation of America, *Best Practices for Identity Theft Services, Version 2.0* (Washington, D.C.: Nov. 17, 2015).

[24] Once exposed, there is no time limit on the potential for identity thieves to use such information to commit fraud.

[25] Massachusetts enacted a law in 2019 that requires companies that experience a breach in which Social Security numbers are disclosed, or reasonably believed to have been disclosed, to offer credit monitoring services at no cost for at least 18 months, or 42 months if the company is a consumer reporting agency, among other things.

the effectiveness of an identity theft provider's services when selecting a vendor to provide such services. Rather, they consider other selection factors, including price, reputation, capacity to respond quickly to large-scale breaches, and ability to provide comprehensive post-breach services, such as complying with statutory notification requirements. But companies that purchase identity theft services may be in a position to obtain more detailed information from potential providers than is publicly available to consumers.

Views of Experts Varied, but Most Said Identity Theft Services Have Limitations and Would Not Address All Data Breach Risks

In the absence of independent evidence of the effectiveness of identity theft mitigation options, we interviewed representatives and reviewed consumer education materials, working papers, and articles from academic, consumer, industry, and government entities. No one solution can protect against the full range of risks to individuals whose personal information was exposed in a data breach, based on our review of documentation and the views of academic, consumer, government, and industry experts. We obtained perspectives on the value of options available to consumers. The following summarizes key observations:

Identity Theft Services

Representatives of 9 of the 10 consumer groups we interviewed generally viewed credit or identity monitoring (or both) to be of limited value. However, one consumer group representative noted that identity monitoring might be useful in circumstances in which Social Security numbers were compromised. In addition, a few consumer group representatives indicated that consumers could consider signing up for such services if they are offered for free. If identity theft services are not free, FTC and CFPB consumer education materials recommend that consumers consider the benefits and limitations of such services and compare them to free or low-cost options before signing up. A few consumer groups and one academic highlighted that consumers may not fully understand the limitations of signing up for identity theft services. A few consumer group representatives and one industry and state government representative cautioned that free services may be offered for only 1 or 2 years; exposed information can be used for identity theft or other harms over a much longer period. For example, in 2017, we reported that nation-state actors that steal consumer data as part of their espionage activities can wait much longer than a private identity thief to use compromised information (if at all), according to one identity theft service provider. In addition, CFPB consumer information and a few consumer group representatives noted that consumers should be aware that some services may try to charge consumers after the free period ends.

Some consumer group and one industry representatives also said that the value of one feature of identity monitoring—dark web monitoring—is unclear. One representative said that there is nothing new that consumers can do once they learn their information was found on an illicit website. Rather, they must continue to monitor their accounts as they already should have been doing. In addition, one consumer group representative indicated that these services may provide consumers with a false sense of security.

Experts we interviewed for our 2017 report said that identity restoration in particular could be helpful to consumers. FTC staff and one consumer group representative we interviewed said that one-on-one assistance can be helpful. Identity restoration typically is included with other identity theft services rather than offered as a stand-alone service. However, the level of service provided in identity restoration can vary substantially—some providers offer individualized hands-on assistance, while others largely provide self-help information that is of more limited value. In our 2017 report, we also found that another feature of identity theft services, identity theft insurance, may provide minimal benefits for consumers. More details about identity theft insurance appear later in this chapter.

Options to Prevent Fraud or Harm Unrelated To Credit Accounts

Consumers have limited options to mitigate risks of other harms from data breaches, such as medical identity theft and identity theft tax refund fraud. Commercial identity theft services, credit freezes, and fraud alerts do not directly address these risks. Some consumer, government, and industry representatives cited self-monitoring as a way for consumers to be on the alert for these other types of fraud.

Consistent with our 2017 report, identity theft service providers we interviewed generally indicated that their products and services do not directly monitor for these types of fraud. However, two noted that they would assist with any identity restoration involving medical identity theft, tax refund fraud, or government benefits fraud (such as fraudulently redirecting Social Security benefits). Identity theft services also may address these types of fraud indirectly—for example, detecting a fraudulent change of address can prevent sensitive health insurance information from being redirected to the fraudster. A few consumer groups said that consumers may not understand which risks commercial identity theft services address. Additionally, we reported in 2017 that identity theft services do not address non-financial harms, such as emotional distress, embarrassment, and harm to one's reputation. For example, a House Committee report on the OPM data breaches noted that the information stolen from background investigations included some of the most intimate and potentially embarrassing aspects of a person's life, such as mental health history, misuse of alcohol or drugs, or problems with gambling.[26] Identity theft services also may be of limited value in cases of nation-state espionage. For example, in 2017, we reported that when the source of the data breach appears to be a nation state (as opposed to a private party), the risk of the information being sold for monetary purposes is likely to be lower, according to an FTC representative.

Importance of Data Security

In the view of some experts, entities such as the federal government and private companies that hold consumer data have a responsibility to protect

---

[26] The OPM Data Breach: How the Government Jeopardized Our National Security for More than a Generation, House Committee on Oversight and Government Reform, 114th Congress (Sept. 7, 2016).

those data.[27] A few experts said that the burden should not be on consumers to protect data they do not control. Except in certain circumstances, companies are generally not required to be transparent about the consumer data they hold or how they collect, maintain, use, and secure these data.[28] Identity theft service providers may contract with third parties such as consumer reporting agencies or with third-party identity monitoring providers, such as dark web monitoring services. Moreover, one consumer group representative noted that identity monitoring services require consumers to provide additional personal information to enroll—which also could be compromised if the service provider's information were breached.

Finally, consumer group and government researchers we interviewed suggested other options that entities can (or already) use to address risks of harm. For example, one government researcher noted that financial institutions have started to use multifactor authentication and other technologies that can help institutions verify a consumer's identity and thus help prevent fraud. Multifactor authentication involves first logging into an online account using the traditional username and password, and then the institution sending a verification code to a mobile phone or e-mail address that the consumer must enter as part of the log-in process. In addition, one researcher noted that some institutions have started to use facial recognition technology, or to ask an account holder to provide answers to questions such as the size of the account holder's last deposit.[29] Other biometric technologies such as fingerprint recognition on mobile phones, or one-time passcodes that are synced with financial institutions' websites, also can help, according to one researcher and one consumer group

---

[27] We previously reported on data security and data protection at entities that store sensitive personal information. See GAO, Data Protection: Actions Taken by Equifax and Federal Agencies in Response to the 2017 Breach, GAO-18-559 (Washington, D.C.: Aug. 30, 2018); and Consumer Data Protection: Actions Needed to Strengthen Oversight of Consumer Reporting Agencies, GAO-19-196 (Washington, D.C.; Feb. 21, 2019).

[28] See GAO, Information Resellers: Consumer Privacy Framework Needs to Reflect Changes in the Technology and the Marketplace, GAO-13-663 (Washington, D.C.: Sept. 25, 2013); and Personal Information: Key Federal Privacy Laws Do Not Require Information Resellers to Safeguard All Sensitive Data, GAO-06-674 (Washington, D.C.: June 26, 2006).

[29] Facial recognition technology is one of several biometric technologies that identify individuals by measuring and analyzing their physiological or behavioral characteristics.

representative. Other strategies can focus on reducing the riskiness of breaches by making information less useful for purposes of committing identity theft. For example, one researcher noted that organizations could encrypt data or use tokens so static account numbers could not be used on their own. There is no single solution to address all risks of harm, based on our review of documentation and the views of academic, consumer, government, and industry experts.

Consumers Can Use Free Credit Freezes and Fraud Alerts to Effectively Prevent New-Account Fraud

A credit freeze is the only consumer option that can prevent one type of identity theft-related fraud, and recent federal legislation made credit freezes free and easier to place or lift. This option is effective because it restricts potential creditors from accessing a consumer's credit report to open a new account until the consumer asks the nationwide consumer reporting agency to remove or temporarily lift the freeze. In contrast, identity theft services and self-monitoring detect or remediate identity theft after it has occurred, but do not prevent the fraud from occurring in the first place. We interviewed representatives, or reviewed the consumer education or informational materials, of consumer, industry, and government entities and found that almost all of them included credit freezes on credit reports as a useful consumer option to protect against identity theft.

More specifically, the Economic Growth, Regulatory Relief, and Consumer Protection Act, which took effect on September 21, 2018, required the three nationwide consumer reporting agencies (Equifax, Experian, and TransUnion) to make placing and lifting freezes free and specifies that the agencies must place a freeze within 1 business day, and lift it within 1 hour, of receiving a telephone or electronic request (see fig.

1).[30] Consumers must contact each of the three agencies individually and request the freeze. Consumers obtain a PIN from each company, which enables them to lift or remove a freeze at a later date. Before the 2018 act, consumers typically had to pay $5-$10 per agency to place a credit freeze. Some experts had noted cost and inconvenience as some of the limitations to a credit freeze. The new law addresses these concerns to some degree by making credit freezes free and requiring these consumer reporting agencies to lift freezes expeditiously on request.

While the new law removed some barriers to placing credit freezes, others still exist and the freezes have some limitations. For example, consumers still have to lift a freeze before applying for a loan or new credit account and need to place or remove a freeze at each consumer reporting agency separately, which could cause delays for consumers actively shopping for a home, car, or other purchase requiring the extension of credit. Two consumer groups said that there is confusion about how the law would affect minor children. (Under the new law, credit freezes only can be placed on behalf of children under age 16, but not minors ages 16 and 17—who must place freezes themselves).

Moreover, as the new law only applies to the three nationwide consumer reporting agencies, credit freezes do not protect against new-account fraud resulting from the use of credit reports from other consumer reporting agencies. For example, one consumer group recommended that consumers place a fourth freeze with the National Consumer Telecom and Utilities Exchange—a consumer reporting agency that maintains credit reports that telecommunications or utilities companies may use to check the creditworthiness of consumers interested in opening phone or utility accounts. The law also permits insurance companies and employers to continue to access credit reports even after they are frozen, among other exceptions.

---

[30] *See* Pub. L. No. 115-174, §301, 132 Stat. 1296, 1326 (2018) (codified at 15 U.S.C. § 1681c–1(i)).The Congressional Budget Office estimated that about 0.3 percent of Americans with credit reports have frozen their credit. Congressional Budget Office, *Congressional Budget Office Cost Estimate, S. 2155, Economic Growth, Regulatory Relief, and Consumer Protection Act* (Washington, D.C.: March 5, 2018).

Selected 2018 Federal Legislative Changes

Free Credit Freezes

- Individuals can place freezes on their credit reports for **free** at the three nationwide consumer reporting agencies—Equifax, Experian, and TransUnion.

Freeze within 1 Business Day

- Consumer reporting agency must place the freeze within **1 business day** of receiving a phone or electronic request (or within **3 business days** of a mailed request).

Unfreeze within 1 Hour

- Consumer reporting agency must remove a freeze within **1 hour** of receiving a phone or electronic request (or within **3 business days** of a mailed request).

Why Credit Freezes Can Be Helpful:

- Prevents new accounts from being opened, in situations where review of a credit report is required.
- Guardians can place credit freezes for minor children (under age 16) or adults who are incapacitated.
- Cited by experts as one of the most effective tools for preventing new-account fraud.
- Relatively convenient: freezes can be managed electronically (or through other methods) and can be removed temporarily for a specified duration.

Consumers Should Be Aware:

- Consumers must request a freeze at each of the three agencies separately.
- Could still cause delays in approval of loans or other credit applications, especially if consumer forgets or loses the personal information number (PIN) the agencies give to consumers to unfreeze their credit reports.
- Freezes do not prevent fraud on existing accounts (for example, the use of a stolen credit card number to make charges on a credit card).
- Does not prevent other types of harm, such as tax refund or medical identity fraud.
- Not all access to credit reports is frozen (for example, still allowed for insurance underwriting and employment background checks).
- Credit reports at agencies other than Equifax, Experian, and TransUnion will not be frozen (for example, those used to open utility accounts).

Source: GAO analysis, Federal Trade Communication, Consumer Financial Protection Bureau, and consumer and industry organizations. | GAO-19-230.

Figure 1. Overview of Credit Freezes.

One general limitation of credit freezes is that they do not protect against new-account fraud in cases in which credit reports are not used to verify a consumer's creditworthiness. Furthermore, credit freezes do not protect against existing-account fraud, such as fraudulent credit card charges, or certain other types of fraud, such as identity theft tax refund fraud or synthetic identity fraud using elements of individuals' identity information.

While experts with whom we spoke across industry, government, and consumer groups generally believed credit freezes to be an effective tool in preventing new-account fraud, some consumer and industry experts indicated that fraud alerts also can be a good alternative for consumers. Unlike a credit freeze, a fraud alert still allows companies to access an individual's credit report for the purpose of opening a loan or credit account. Fraud alerts notify companies requesting the reports that the individual may have been a victim of identity theft. The alerts require companies to verify consumers' identities before they issue credit to a consumer.[31] Fraud alerts therefore can make it harder for an identity thief to open accounts in a consumer's name. Moreover, fraud alerts are easier to place than credit freezes, as consumers only need to contact one of the three nationwide consumer reporting agencies to place a fraud alert (that agency is then obligated to contact the other two on the individual's behalf). The Economic Growth, Regulatory Relief, and Consumer Protection Act extended the period of an initial fraud alert from 90 days to 1 year.[32]

However, fraud alerts do not restrict access to consumers' credit reports the way freezes do. Therefore, some consumer group and industry representatives noted that consumers should be aware that a fraud alert may not offer as strong a protection as a credit freeze does. We did not find any data or analysis on the effectiveness of fraud alerts compared to credit freezes or monitoring options.

---

[31] *See* 15 U.S.C. § 1681c-1(h)(1)(B). Consumers can request an initial fraud alert, extended fraud alert, or active duty alert at no cost with any one of the three nationwide consumer reporting agencies, which automatically must notify the other two. *See* 15 U.S.C. § 1681c-1(a)-(c). An initial fraud alert stays on the victim's credit file for 1 year after which the consumer may place another fraud alert. An extended fraud alert, which lasts for 7 years, is available to victims of identity theft who have filed a formal identity theft report with one of the three agencies. Active duty alerts, which last for 1 year, are available to deployed military service members.

[32] See Pub. L. No. 115-174, §301(a)(1), 132 Stat. 1296, 1326 (2018) (codified at 15 U.S.C. § 1681c–1(a)(1)(a)). Additionally, the act mandated that FTC issue a rule regarding the law's requirement that the nationwide consumer reporting agencies provide a free electronic credit monitoring service that would notify active-duty military members within 24 hours of any "material" additions or modifications to their credit files. § 302(d)(1).

One consumer group told us that it recommends that after a data breach consumers first place a fraud alert, because it requires contacting only one of the three nationwide consumer reporting agencies, and then follow up by placing a credit freeze at the three agencies.

The three nationwide consumer reporting agencies also offer a product called a credit lock that is functionally similar to a credit freeze in that it restricts access to an individual's credit report. Credit locks do not require consumers to use a PIN and consumers can turn access to credit reports on or off through an application on their mobile phone. However, credit locks are not subject to the same federal requirements regarding the placement and removal of freezes and therefore do not offer the same degree of protection to consumers.[33] Instead, credit locks are private products subject to the consumer reporting agencies' terms and conditions, which could change. A credit lock is in place only as long as the individual subscribes to an agency's service, but a credit freeze remains in place until the consumer chooses to remove it. Finally, consumers may be charged a fee to place a credit lock, whereas credit freezes can now be placed for free.

Factors Consumers Can Consider When Assessing Options after Data Breaches

Based on our interviews and review of consumer education materials and our 2017 report, we identified four factors that consumers can consider in deciding which options are best for them in responding to a breach of their personal information:

---

[33] The Fair Credit Reporting Act protects the accuracy and confidentiality of personal information collected or used for eligibility determinations for such purposes as credit, insurance, or employment. As such, it regulates the collection and use of consumers' personal information by consumer reporting agencies. Moreover, the Fair Credit Reporting Act requires information on credit freezes to be included in any notice of consumer rights required under the act. *See* 15 U.S.C. § 1681c-1(i)(5).

- *Prevention.* Consumers can consider the extent to which an option might prevent fraud. For example, because credit freezes block all access to an individual's credit report, by definition they are effective in preventing new-account fraud where credit reports are used as part of the account-opening process. Identity theft services do not prevent fraud, but detect suspicious activity or help restore identities after identity theft.
- *Cost.* Consumers can consider the cost of a service. For instance, consumers can consider whether to pay for commercial identity theft services if they believe the value of the service outweighs the effort of monitoring their accounts on their own. In addition, they may consider that credit freezes now are available for free.
- *Convenience.* Consumers may consider the convenience of a service. For example, while consumers can monitor their own credit reports and accounts, some might prefer not to or may be limited in their ability to do so. In addition, technologies offered through financial institutions that automatically alert customers to any transactions involving their accounts can be a convenient, no-cost way for consumers to monitor their accounts.
- *Type of information at risk.* Finally, several experts from consumer and industry organizations indicated that the type of option that might be beneficial would depend on the type of information at risk. For example, one consumer group representative noted that if a credit card number were stolen, an identity monitoring service that monitored the dark web for Social Security numbers might not be needed. Furthermore, consumers should consider that credit monitoring will be of limited effectiveness in alerting them to misuse of an existing credit account—which is more common than fraud related to setting up new accounts.[34] For more information on consumers' options, see appendix II.

---

[34] As noted previously, for 85 percent of identity-theft victims in 2016, the most recent incident involved misuse or attempted misuse of one type of existing account, such as a credit card or bank account. Only 1 percent of those 16 or older experienced the opening of a new account or other misuse of personal information apart from misuse of an existing account.

Federal Agencies Provide Assistance to Consumers Affected by Data Breaches and Identity Theft

Among federal agencies, FTC serves as a primary source for free assistance (including online resources, educational outreach, and customized assistance through IdentityTheft.gov) to consumers on ways to respond to data breaches, identity theft, and related harm. Approximately 13 percent of those affected by the 2015 OPM breaches used credit and identity monitoring and identity restoration services that OPM offered them and a fraction of a percent made identity theft insurance claims (the payouts for which averaged $1,800). Data we assessed for this chapter support a 2017 recommendation we made to the Office of Management and Budget (OMB) to revise guidance to federal agencies about responding to data breaches and one to Congress to consider permitting agencies to determine appropriate levels of identity theft insurance offered after data breaches.

FTC Is Primary Provider of Federal Assistance to Consumers Affected by Data Breaches and Identity Theft

Federal Trade Commission

FTC, as a primary source for assistance to consumers on issues related to data breaches and identity theft, provides guidance and assistance through its website and through conferences and workshops.[35]

See Department of Justice, Bureau of Justice Statistics, *Victims of Identity Theft, 2016* (January 2019).

[35] The Identity Theft and Assumption Deterrence Act of 1998 establishes FTC as the central clearinghouse for identity theft victim complaints and directs FTC to provide consumer education to identity theft victims. *See* Pub. L. No. 105-318, § 5(a), 112 Stat. 3007, 3010 (codified at 18 U.S.C. § 1028 note). More recently, an Executive Order included a provision calling for federal agencies to centralize identity theft information and resources at FTC's website, IdentityTheft.gov, and through the general FTC website. Exec. Order No. 13681, 78 Fed. Reg. 63491, 63492 (Oct. 23, 2014). FTC's primary legal authority comes from

Online and Printed Resources

FTC's home page includes links to identity theft-related resources, including information about key options consumers can consider to help them mitigate identity theft risks and other harms, and a link to IdentityTheft.gov (discussed later in this section). FTC updates the information regularly, such as after large-scale data breaches.

Outreach

FTC maintains relationships with state government, law enforcement, and community and consumer organizations, through which it conducts outreach about how to respond to exposure or loss of personal information and identity theft mitigation. For example, FTC collaborated with the International Association of the Chiefs of Police to update the association's model policy for identity theft to include referral information for IdentityTheft.gov. FTC also has held webinars, conferences, and workshops on topics related to data breaches and identity theft for groups including government officials, nonprofits, and the general public.

Customized Assistance (IdentityTheft.gov)

FTC provides information and customized assistance through IdentityTheft.gov to individuals whose information was lost or stolen or who experienced identity theft or other harm, such as tax refund fraud. During fiscal year 2018, IdentityTheft.gov received almost 2 million unique visitors. The website in its current form has been in place since January 2016 and offers the following types of assistance:

---

section 5 of the Federal Trade Commission Act, which prohibits unfair or deceptive acts or practices in the marketplace. *See* 15 U.S.C. § 45. FTC has authority to enforce sector-specific laws, including the Truth in Lending Act, the CANSPAM Act, the Children's Online Privacy Protection Act, the Equal Credit Opportunity Act, the Fair Credit Reporting Act, the Fair Debt Collection Practices Act, and the Telemarketing and Consumer Fraud and Abuse Prevention Act. As directed by Congress, FTC also has authority to issue rules that regulate specific areas of consumer privacy and security. For example, FTC's Red Flags Rule requires financial institutions and certain creditors to have programs to identify, detect, and respond to patterns, practices, or specific activities that could indicate identity theft.

- *Steps to take after identity theft. IdentityTheft.gov* provides individual victims with step-by-step instructions to resolve specific problems. From January 2016 (when FTC launched the current version of IdentityTheft.gov) through October 1, 2018, approximately 700,000 individuals set up and activated accounts on the website to help them recover from identity theft. Individuals who set up accounts can indicate what kind of information was stolen and what kind of adverse event they experienced. The site helps users generate pre-filled letters, affidavits, and forms to send to consumer reporting agencies, businesses, debt collectors, and IRS, as appropriate. For example, individuals who fill out an Identity Theft Report affidavit can use this chapter instead of filing a police report to request extended 7- year fraud alerts (available to identity theft victims) on their credit reports. In addition, individuals who experienced tax refund fraud can fill out a form on IdentityTheft.gov that is then submitted directly to IRS. An individual who experienced credit card fraud would be advised to take different steps than one who experienced fraud related to utility bills or medical insurance.
- *Steps to take after data breaches or loss of personal information.* IdentityTheft.gov/databreach provides checklists and suggestions for people whose personal information was lost or exposed but has not yet been misused.

FTC also maintains an online chat function and telephone number for those who need additional assistance. For complex cases, FTC staff may refer individuals to the Identity Theft Resource Center, a nonprofit organization.

We found that in developing and updating the website, FTC followed some key practices for consumer education planning.[36] One key practice we identified was consulting with stakeholders. According to FTC staff we interviewed and documentation we reviewed, FTC obtained feedback from

[36] In GAO-08-43, we describe key practices for conducting consumer education identified by an expert panel that we convened.

stakeholders such as law enforcement agencies and community organizations in developing IdentityTheft.gov. Another key practice we identified was assessing users' needs. FTC conducted usability testing to ensure the site's features were easy to use. FTC staff also told us that after receiving user feedback, they made it easier for users to set up an account. FTC also made changes to IdentityTheft.gov—such as incorporating the ability to auto-generate forms—to implement a 2014 Executive Order calling for federal agencies to centralize identity theft information at the website.[37] Furthermore, in January 2018, FTC implemented a new function that allows users who report identity theft tax refund fraud to file reports directly with IRS. Since its launch in early 2018 through October 1, 2018, almost 22,000 IRS Identity Theft Affidavits (IRS Form 14039) were submitted to IRS through IdentityTheft.gov. In general, experts across consumer, government, and industry organizations and identity theft service providers we interviewed expressed the view that IdentityTheft.gov is a valuable or user-friendly resource.

Other Federal Agency Resources

Other federal agencies provide assistance to consumers on topics related to identity theft, including CFPB, the Department of Justice, IRS, and the Social Security Administration.[38]

CFPB

CFPB enforces, supervises for compliance with, and issues regulations to implement the federal consumer financial laws that address certain firms' and financial institutions' practices, which may include data security. A few of these laws and regulations contain provisions that can

---

[37] Exec. Order No. 13681, 78 Fed. Reg. 63491 (Oct. 23, 2014).

[38] For example, see Department of Justice, Identity Theft, https://www.justice.gov/criminal-fraud/identity-theft/identity-theft-and-identity-fraud, which we accessed on November 19, 2018. Also see Internal Revenue Service, *Taxpayer Guide to Identity Theft,* https://www.irs.gov/newsroom/taxpayer-guide-to-identity-theft, accessed on November 19, 2018; *Data Breach: Tax-Related Information for Taxpayers*, May 2018; and *Identity Protection: Prevention, Detection, and Victim Assistance,* July 2018. See Social Security Administration, *Identity Theft and Your Social Security Number*. These agencies were outside the scope of our review for this report, but we previously reported on some related topics (for instance, see GAO-18-418).

help protect the personal information of consumers.[39] CFPB also offers consumer education resources.

Similarly to FTC, CFPB included information about how consumers can address risks related to exposure of personal information and recover from identity theft in the bureau's overall consumer education activities. CFPB provides consumer education materials related to data breaches and identity theft through its blog and its financial education resource, "Ask CFPB." CFPB also maintains relationships with external groups, such as librarian networks. CFPB provides links to FTC resources about data breaches and identity theft topics on its website, so as not to duplicate efforts, according to CFPB staff.[40] The two agencies also have coordinated some efforts. FTC and CFPB published a jointly produced blog post on September 21, 2018, the date the new free credit freeze and 1-year fraud alert provisions took effect. Such coordination is consistent with the 2014 Executive Order, which designated FTC as a centralized source of information about identity theft across the federal government.

Staff of both agencies said that in developing new resources, they monitor information from a variety of sources, including consumer complaints, news and social media, and reports from other government entities, law enforcement, or nongovernmental stakeholders.

Other Federal and State Agencies

IRS and the Social Security Administration provide some assistance to consumers for specific types of identity theft. For example, as noted previously, IRS provides some taxpayers with PINs if they are victims of

---

[39] These include sections 502 through 509 of the Gramm-Leach Bliley Act, sections 1031 and 1036 of the Dodd-Frank Wall Street Reform and Consumer Protection Act concerning unfair, deceptive, or abusive acts and practices, the Fair Credit Reporting Act, and the Fair Debt Collection Practices Act. *See* Pub. L. No. 111-203, § 1002(12), 124 Stat. 1376, 1957 (2010) (codified at 12 U.S.C. § 5481(12)). CFPB has no authority under section 501(b) of the Gramm-Leach Bliley Act, which required FTC and other agencies (but not CFPB) to establish standards for financial institutions on administrative, technical, and physical safeguards to ensure the security and confidentiality of customer records and information, to protect against anticipated threats or hazards to the security or integrity of such records, and protect against unauthorized access to or use of such records or information that could result in substantial harm or inconvenience to any customer. *See* 15 U.S.C. § 6801(b).

[40] Other federal agencies' websites also provide links to FTC resources, including those of OPM, IRS, the Department of Justice, and the Social Security Administration.

identity theft tax refund fraud. In addition, states enforce laws and regulations and provide consumer education resources and assistance to consumers at risk of identity theft and other harms as a result of data breaches. For example, the Illinois Attorney General's office maintains a call-in number for victims of identity theft, and the Colorado Bureau of Investigation can assist residents with identity theft issues.

Few People Used Identity Theft Services OPM Provided, Very Few Made Insurance Claims, and Payouts Received Were Low

OPM offered identity theft services to approximately 22.1 million individuals whose personal information was compromised during the 2015 data breaches at OPM.[41] Personnel records or OPM systems containing information from the background investigations of current, former, and prospective federal employees and other individuals were breached. The services, offered at no cost to affected individuals, included credit monitoring, identity monitoring, identity restoration services, and identity theft insurance.[42] To receive credit and identity monitoring services, affected people have to enroll with the identity theft service provider with which OPM contracted, but identity theft insurance and identity theft restoration services are available to the entire affected population whether or not they enroll.

Few affected individuals have used the services. According to data from OPM, as of September 30, 2018, close to 3 million, or 13 percent, of individuals affected by the 2015 incidents had made use of the services. As seen in figure 2, the great majority of enrollments occurred in the months

[41] The federal contracts awarded in 2015 for the OPM data breaches provided for federal agencies to receive information about the identity theft services delivered by contractors, such as information about call-center wait times and the number and status of identity restoration cases.

[42] The identity restoration and insurance provided to consumers covers all incidents of identity theft that occur during the coverage period, regardless of the source.

immediately following notification of the breach.[43] OPM staff said that the spike in enrollments in July and August 2016 likely was due to the follow-up mailing that OPM sent to approximately 10 percent of affected individuals whose mailing addresses were incorrect in the original mailing of notifications.[44]

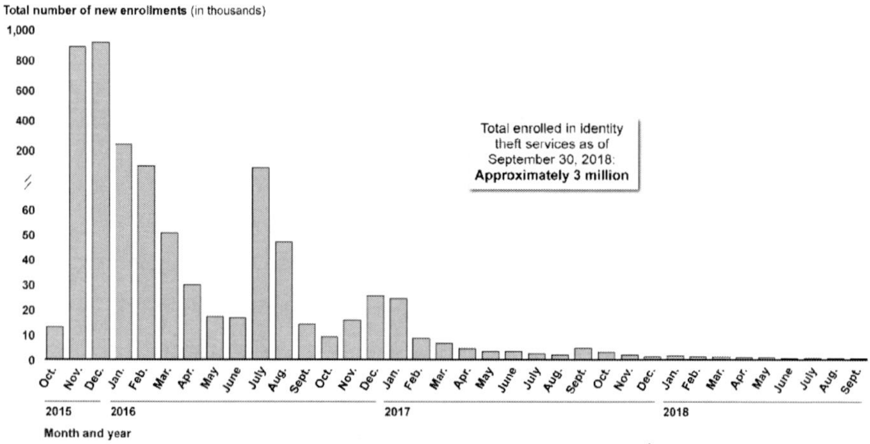

Source: GAO analysis of Office of Personnel Management data. | GAO-19-230.

Notes: The Office of Personnel Management (OPM) offered identity theft services to approximately 22.1 million individuals whose sensitive personal information was compromised during the 2015 data breaches of OPM personnel records or OPM systems containing information from the background investigations of current, former, and prospective federal employees and other individuals. The services included credit and identity monitoring, identity restoration services, and identity theft insurance. According to OPM staff, the spike in enrollments in July and August of 2016 likely was due to the follow-up mailing that OPM sent to approximately 10 percent of affected individuals whose mailing addresses were incorrect in the original September 2015 notification.

Figure 2. Enrollment in Identity Theft Services Offered by Office of Personnel Management, October 2015–September 2018.

---

[43] OPM data show that of the approximately 3 million individuals enrolled in identity theft services through September 30, 2018, there were about 2.7 million adults and 350,000 minors. See GAO-17-254 for more information.

[44] OPM worked with the Department of Defense and a commercial address validation service to identify the correct addresses and sent out notifications to these individuals in June 2016.

In addition, according to OPM-reported data we reviewed, of the 3 million individuals who used the services, about 1 percent made identity restoration requests and a fraction of 1 percent submitted insurance claims. According to data we reviewed, approximately 27,000 identity restoration cases had been resolved as of September 30, 2018. In addition, 61 insurance claims (of 81 submitted) had been paid, totaling $112,000, with an average payout of $1,800.[45]

Since 2015, OPM has obligated approximately $421 million for identity theft services and as of November 30, 2018, OPM paid out approximately$361 million of the obligated funds.[46] OPM is required to provide identity theft services through September 2026.[47] The contract to provide these services on behalf of OPM expired in December 2018; OPM re-competed and awarded a single contract that month to ID Experts, the company that had been providing these services.[48]

After the OPM breaches in 2015, OPM provided federal employees and other affected individuals with information and guidance about their options in mailed letters and on its website. On its website, OPM developed a Cybersecurity Resource Center and included background about the breaches and who was affected; instructions for how to enroll in identity theft services; and a Frequently Asked Questions webpage that included links to FTC resources, including IdentityTheft.gov. OMB's 2017 policy guidance to federal agencies, including OPM, states that agencies

---

[45] As stated previously, all incidents are covered regardless of the source.

[46] In GAO-17-254, we found that OPM's breach-response policies and procedures did not specifically address identity theft services, which could hinder informed decision-making by the agency on the appropriate services, if any, to offer affected individuals. Therefore, we recommended that OPM incorporate criteria and procedures for determining whether to offer identity theft services into the agency's policy on responding to data breaches. We also found that OPM had not adequately documented how it made its decisions about its 2015 breaches. We recommended that the agency implement procedures to help assure that significant decisions on the use of identity theft services would be appropriately documented. In September 2017, OPM implemented both recommendations.

[47] *See* Consolidated Appropriations Act, 2017, Pub. L. No. 115-31, § 633, 131 Stat. 135, 376.

[48] According to OPM staff, the contract was competed through a blanket purchase agreement bidding process. A blanket purchase agreement is a contracting vehicle that agencies are encouraged to use in order to easily access and acquire qualified providers on prenegotiated prices for services. The General Services Administration has established an identity theft services blanket purchase agreement, which includes selected identity theft service providers.

should determine appropriate information to provide to affected individuals and review breach responses annually. Consistent with that guidance, OPM's September 2017 Breach Response Plan calls for the agency to review its breach response plan annually, including to reinforce or improve training and awareness. In December 2018, OPM updated its website to incorporate changes in the cost of credit freezes and duration of fraud alerts resulting from new legislation we discussed earlier.

OMB Has Not Revised Post-Data Breach Guidance to Agencies and Insurance Coverage Amount for Identity Theft Insurance Remains High

Data we assessed for this chapter support a 2017 recommendation we made to OMB and a matter for congressional consideration, both of which have not yet been implemented.[49] In our March 2017 report, we found that OMB policy guidance for federal agencies on how to prepare for and respond to data breaches did not address how agencies might assess the effectiveness of identity theft services relative to lower-cost alternatives.[50] For example, the guidance did not discuss whether identity theft services would be preferable to alternatives (such as fraud alerts, credit freezes, or the agency conducting its own database monitoring). We concluded that the guidance might not fully reflect the most useful and cost-effective options agencies should consider in response to a breach—contrary to OMB's risk-management and internal control guidance calling on federal leaders to improve effectiveness and efficiency. Therefore, we recommended that OMB conduct an analysis of the effectiveness of identity theft services relative to alternatives, and revise its guidance to

---

[49] See GAO-17-254.

[50] OMB Memorandum M-17-12. In this memorandum, OMB defines personally identifiable information as information that can be used to distinguish or trace an individual's identity, either alone or when combined with other information linked or linkable to a specific individual. OMB was directed to update this guidance by the October 2015 Cybersecurity Strategy and Implementation Plan for the Federal Civilian Government. See Office of Management and Budget, *Cybersecurity Strategy and Implementation Plan for the Federal Civilian Government*, M-16-04 (Washington, D.C.: Oct. 30, 2015).

federal agencies in light of the analysis. In oral comments on a draft of the 2017 report, staff from OMB's Office of Information and Regulatory Affairs said that our draft recommendation to OMB on expanding OMB's guidance to federal agencies would benefit from greater specificity, and we revised this recommendation to provide greater clarity.

We contacted OMB several times between May 2018 and early March 2019 to update the status of this recommendation but as of March 2019, OMB had not responded with an update.[51] In our current review, we found that information on the effectiveness of various consumer options continues to be limited. We also found that some free and low-cost alternatives to free or fee-based identity theft services can prevent or more directly address new account fraud and some options consumers can take on their own have become less burdensome. Therefore, we stand by this recommendation.

In addition, as noted previously in this chapter, the identity theft insurance that OPM offered to affected individuals resulted in few insurance claims, and the amounts claimed have been small. These data are consistent with the findings of our 2017 report—which reported that the number and dollar amount of claims for identity theft generally were low.[52] They also reinforce our conclusion that the $5 million per-person coverage limit mandated by Congress likely was unnecessary and might impose costs without providing a meaningful corresponding benefit. Specifically, we noted that $5 million in coverage would increase federal costs

---

[51] GAO-17-254 also recommended that OMB explore options to address the risk of duplication in federal provision of identity theft services in response to data breaches, and take action if viable options were identified. In November 2018, OMB staff told us this recommendation would be updated as part of OMB's response to our request for the status of recommendations in our annual report on opportunities to reduce fragmentation, overlap, and duplication, reduce costs, and increase revenue, for the government. See GAO, *2018 Annual Report: Additional Opportunities to Reduce Fragmentation, Overlap, and Duplication and Achieve Other Financial Benefits*, GAO-18-371SP (Washington, D.C.: Apr. 26, 2018).

[52] GAO-17-254. We found identity theft insurance is limited in covering direct financial losses—that is, money that was stolen. Instead, the insurance generally reimburses consumers for out-of-pocket expenses they incur related to the process of restoring their identity and credit records. While the overall coverage limit for policies can be quite high (around one million dollars), the process of resolving identity theft typically does not require significant expenses, according to many providers with which we spoke and two consumer groups.

unnecessarily, likely mislead consumers about the benefit of the product, and create unwarranted escalation of coverage amounts in the marketplace.

Therefore, we reiterate the matter for congressional consideration we made in our March 2017 report: in the event that Congress again requires an agency to provide individuals with identity theft insurance in response to a breach, it should consider permitting the agency to determine the appropriate level of that insurance.

AGENCY COMMENTS

We provided a draft of this chapter to CFPB, FTC, and OPM. The agencies provided technical comments, which we incorporated as appropriate.

We are sending copies of this chapter to the appropriate congressional committees, the Director of CFPB, the Chair of FTC, and the Acting Director of OPM.

Anna Maria Ortiz
Acting Director, Financial Markets and Community Investment

APPENDIX I: OBJECTIVES, SCOPE, AND METHODOLOGY

This chapter examines (1) information and expert views about the effectiveness of options consumers can use to prevent or address the risks resulting from data breaches; and (2) federal assistance available to help consumers understand these options, including the status of one matter for congressional consideration and one recommendation relating to these issues in our 2017 report.[53]

[53] See GAO, Identity Theft Services: Services Offer Some Benefits but Are Limited in Preventing Fraud, GAO-17-254 (Washington, D.C.: Mar. 30, 2017).

To address the first objective, we conducted a literature review to identify any studies or independent research on the effectiveness of various options consumers have for mitigating data breach harms, consumer attitudes and behavior following data breaches, and identity theft and other harm to individuals from exposure of personal information. We searched databases of scholarly publications and other sources for work generally published within the last 5 years. Examples of databases searched include ProQuest, EconLit, Policy File Index, and SciTech Premium Collection. We searched for terms including "effective," "data breach," "identity theft," "consumer attitudes," and "consumer behavior" and options such as "credit freeze," "fraud alert," and "credit lock." We also reviewed relevant academic literature to identify additional studies. From these searches, we did not identify any studies that assessed the extent to which commercial identity theft services were effective in preventing or mitigating harm from exposure of personal information. We identified and reviewed 54 studies that appeared in peer-reviewed journals or research institutions' publications and were relevant to consumer attitudes and behavior related to privacy, data breaches, and identity theft.

To ensure the selection of a range of perspectives on the effectiveness of options to mitigate harms, we reviewed the selection of experts and sources in our prior report and our literature review, and updated that selection through additional searches and recommendations from discussions with experts and identity theft service providers and review of relevant literature. We defined experts as those representing consumer and industry policy organizations that have conducted research or taken policy positions on consumers' or entities' options after data breaches; academics who conducted research on relevant topics; and federal and state government staff with specific positions of responsibility in consumer protection or education. We also contacted seven companies that provide identity theft services to consumers.

We interviewed representatives of a nongeneralizable sample of 35 entities in the following categories: academic or independent research institution (4); consumer or privacy research and advocacy (10); industry association, identity theft service provider, or industry consultant (12); and federal or state government (9). We also reviewed relevant consumer education and other materials produced by consumer, government, industry, and other entities. We interviewed academics from Carnegie-Mellon University, RAND Corporation, the University of Maryland, and the University of Rochester. In addition, we interviewed representatives from the following organizations:

- *Consumer or privacy groups:* AARP, Consumer Action, Consumer Federation of America, Consumer Reports, Electronic Privacy Information Center, Identity Theft Resource Center, National Consumer Law Center, Privacy Rights Clearinghouse, U.S. PIRG, and World Privacy Forum.
- *Industry associations or consultants:* American Bankers Association, Consumer Data Industry Association, Property and Casualty Insurers Association of America, National Retail Federation, and Rational 360.
- *Identity theft service providers:* Credit Karma, Equifax, Experian, ID Experts, ID Shield, LifeLock, and TransUnion.[54]
- *Government agencies:* Consumer Financial Protection Bureau (CFPB), Federal Reserve Bank of Philadelphia, Federal Trade Commission (FTC), Office of Personnel Management (OPM), and Offices of the Attorney General of California, Connecticut, Illinois, Massachusetts, and New York.[55]

Throughout this chapter, we use certain qualifiers when describing responses from interview participants and views of entities whose articles and written material we reviewed, such as “few,” “some,” and “most.” We define few as a small number such as two or three. The specific quantification of categories depends on the overall numbers of entities that addressed a specific topic. For example, we may refer to views shared by a

[54] We also reviewed the website of Credit Sesame.

[55] We also reviewed the websites of the Offices of the Attorney General of Florida, Indiana, and Texas and of the Colorado Office of Victims Services.

proportion of the 10 consumer groups we interviewed, or those shared by identity theft service providers.

We also reviewed provisions in the Economic Growth, Regulatory Relief, and Consumer Protection Act, enacted in May 2018, that address credit freezes and fraud alerts (two tools for preventing new-account fraud).[56]

To address the second objective, we reviewed and analyzed documentation and interviewed staff from FTC, CFPB, and OPM. We reviewed and analyzed FTC, CFPB, and OPM consumer education materials including blog posts, online fact sheets, and printed brochures and data on usage of the materials. For example, we analyzed FTC, CFPB, and OPM data and website analytics for their data breach- and identity theft-related web pages. We interviewed FTC and CFPB agency staff about their assistance to individuals and how they measure effectiveness of their efforts. We reviewed documentation and interviewed agency staff about the development, implementation, and assessment of consumer education materials and other resources and assistance. For example, we reviewed materials documenting FTC's outreach to stakeholders and usability testing of IdentityTheft.gov. We compared the activities against a 2014 Executive Order on the security of consumer financial transactions, key practices for consumer education planning we identified in prior work, and federal standards for internal control.[57]

We analyzed data from the company with which OPM contracted to provide identity theft services to the approximately 22.1 million individuals whose information was exposed in the 2015 data breaches. We obtained data on the number of enrollments, the number and size of identity theft insurance claims submitted and paid, and number of identity restoration cases the companies handled. We assessed the reliability of the

---

[56] Pub. L. No. 115-174, § 301(a), 132 Stat. 1296, 1326 (2018) (codified at 15 U.S.C. § 1681c–1(i)).

[57] Exec. Order No. 13681 79 Fed. Reg. 63491(Oct. 23, 2014). See GAO, Digital Television Transition: Increased Federal Planning and Risk Management Could Further Facilitate the DTV Transition, GAO-08-43 (Washington, D.C.: Nov. 19, 2007); and Standards for Internal Control in the Federal Government, GAO-14-704G (Washington, D.C.: Sept, 10, 2014). Also see Office of Management and Budget, Preparing for and Responding to a Breach of Personally Identifiable Information, M-17-12 (Washington, D.C.: Jan. 3, 2017).

data by interviewing agency officials and reviewing documentation about the systems used to store the data. We found the data to be reliable for purposes of this reporting objective. We also reviewed the online guidance OPM provided to affected individuals and assessed the guidance against Office of Management and Budget guidance for agencies following data breaches and OPM's 2017 Breach Response plan.

In addition, for both objectives, we reviewed the evidence gathered and analyzed for the 2017 GAO report (GAO-17-254) and updated the status of the matter for congressional consideration and recommendations made in that report.[58]

We conducted this performance audit from November 2017 to March 2019 in accordance with generally accepted government auditing standards. Those standards require that we plan and perform the audit to obtain sufficient, appropriate evidence to provide a reasonable basis for our findings and conclusions based on our audit objectives. We believe that the evidence obtained provides a reasonable basis for our findings and conclusions based on our audit objectives.

APPENDIX II: WHAT CAN CONSUMERS DO AFTER A DATA BREACH?

Figure 3 below provides information on actions consumers can take to monitor for identity theft or other forms of fraud, protect their personal information, and respond if they have been a victim of identity theft. This information summarizes prior GAO work and comments of academic, consumer organization, industry, and government experts.[59]

---

[58] See GAO, Identity Theft Services: Services Offer Some Benefits but Are Limited in Preventing Fraud, GAO-17-254 (Washington, D.C.: Mar. 30, 2017).

[59] GAO, Identity Theft Services: Services Offer Some Benefits but Are Limited in Preventing Fraud, GAO-17-254 (Washington, D.C.: Mar. 30, 2017).

Prevent Fraud on New Credit Accounts

Consumer Option	How This Option Can Help	Consumers Should Be Aware
Place a credit freeze on credit reports at Equifax, Experian, and TransUnion—the three nationwide consumer reporting agencies.	• Prevents identity thieves from opening new credit accounts in an individual's name—where credit reports are required. • Guardians can place credit freezes for minor children (under age 16) or adults who are incapacitated.	• Consumers must request a freeze at each of the three agencies separately. • Could still cause delays in approval of loans or other credit applications, especially if consumer forgets or loses the personal information number (PIN) the agencies give to consumers to unfreeze their credit reports. • Freezes do not prevent fraud on existing accounts (for example, the use of a stolen credit card number to make charges on a credit card). • Freezes do not prevent other types of harm, such as tax refund or medical identity fraud. • Not all access to credit reports is frozen (for example, still allowed for insurance underwriting and employment background checks). • Credit reports at agencies other than Equifax, Experian, and TransUnion will not be frozen (for example, those used to open utility accounts).
Place a fraud alert at the three nationwide consumer reporting agencies, which lasts 1 year and can be renewed.	• Fraud alerts let businesses know that a consumer may have been a victim of fraud. • Businesses must take extra steps to verify the identity of the individual seeking to open accounts. • Members of the military can place active duty alerts.	• Consumers can request a fraud alert at one of the three agencies and this agency must notify the other two to place the alert. • Victims of identity theft can place extended fraud alerts that last for 7 years. • Fraud alerts still allow access to credit reports. • Businesses that do not use the three agencies will not see the alert.

Source: GAO analysis, Federal Trade Commission, Consumer Financial Protection Bureau, and consumer and industry organizations.

Monitor for Some Types of Fraud on Financial Accounts

Consumer Option	How This Option Can Help	Consumers Should Be Aware
Review free credit reports every 12 months (from Equifax, Experian, and TransUnion) at annualcreditreport.com.	• Can help consumers spot suspicious activity or fraud involving credit accounts.	• Consumers can check one of the three reports every 4 months to improve chances of catching problems throughout the year.
Review bank and other financial account statements regularly or set up free automatic alerts.	• Can alert consumers to suspicious activity on their accounts.	• The availability and features of alerts may vary among financial institutions.
Consider enrolling in credit or identity monitoring services.	• Credit monitoring can alert consumers after the fact that someone may have used their personal information to open a credit account (take out a loan or sign up for a credit card). • Identity monitoring can alert consumers of misuse of personal information or appearance of their information on illicit websites (the "dark web").	• These services do not directly address risks of medical identity theft, identity theft tax refund fraud, or government benefits fraud. • Credit monitoring can spot fraud but generally cannot prevent it, and does not identify fraud on existing or noncredit accounts. • Identity monitoring also cannot prevent fraud. • It is unclear what actions consumers can take once alerted that their information appears on the dark web other than continuing to monitor their accounts. • These services may be part of a package of identity theft services, including restoration services, or identity theft insurance. • Free services that entities that have experienced data breaches may offer to affected consumers vary in the type and level of service and may only last for 1-2 years. Risks can exist for much longer. • Paid services typically cost \$5–\$30 a month.

Monitor for Other Types of Identity Theft or Fraud

Consumer Option	How This Option Can Help	Consumers Should Be Aware
Mobile Phone or Utility Account Fraud		
Review mobile phone and utility bills regularly.	• Can spot suspicious activity on existing accounts.	• Consumers with credit freezes may need to lift them before applying for new utility or phone accounts.
Medical Identity Theft		
Review medical bills and health insurance explanations of benefits.	• Can spot suspicious activity, such as bills or insurance claims for services consumers did not receive.	• Consumers who spot problems can contact fraud departments at health insurers.

Figure 3. (Continued).

Identity Theft Tax Refund Fraud		
File tax returns early.	• Provides less time for a fraudster to file in an individual's name.	• Consumers who experience identity theft tax refund fraud can file affidavits with the Internal Revenue Service (IRS) and through IdentityTheft.gov, and may be eligible to obtain an Identity Protection Personal Identification Number from IRS.
Government Benefits Fraud		
Set up an online account at the Social Security Administration and check it regularly.	• Can spot suspicious activity, such as benefits redirected to another address.	• Other government benefits, such as unemployment insurance, also can be susceptible to identity fraud.

How to Respond after Identity Theft

Consumer Option	How This Option Can Help	Consumers Should Be Aware
Visit identityTheft.gov to set up an account, fill out, and file necessary reports.	• Helps users determine what steps to take depending on the type of information stolen or type of identity theft. • Can generate an Identity Theft Report that can be used to help contact consumer reporting agencies, law enforcement, and other entities. • Can generate an IRS Identity Theft Affidavit (IRS Form 14039) that can be submitted directly to IRS. • Provides information on what companies to contact and how to remove incorrect information.	• The Federal Trade Commission (FTC) also has a telephone help line and online chat feature.
Contact state or local government resources, such as consumer protection help lines or victim services offices.	• Some states and local governments can provide one-on-one assistance.	• States and localities vary in the services offered.
Consider using commercial identity restoration services.	• Can reduce consumer time and effort in dealing with the effects of identity theft, such as by interacting with creditors on the consumer's behalf.	• Service levels can vary significantly among companies. Some provide hands-on assistance, while others largely provide information. • May be included in a package of identity theft services, which may also include credit or identity monitoring or identity theft insurance. Paid services typically cost \$5–\$30 a month and free services may only be offered for 1-2 years.

Figure 3. (Continued).

Protect Personal Information in Other Ways

Consumer Option	How This Option Can Help	Consumers Should Be Aware
Adopt Good Practices for Online Accounts • Protect passwords and do not re-use them. • Use two-factor authentication when offered (for example, entering a one-time code sent to a mobile phone when logging in to an online account). • Choose strong passwords and consider using a software application that helps manage passwords. • Do not click on links in emails or open attachments from unknown senders. • Remember that public WiFi may not be secure.	• Can prevent unauthorized access to online accounts and other data intrusions.	• While personal security practices are important, consumers have limited control over how private entities secure their data.
Protect social media accounts by checking privacy settings, and consider limiting information shared.	• Restricts how much information is visible to strangers and their ability to misuse it.	• Privacy terms and conditions can change, so it is important to check settings periodically.
Do not provide personal information over the phone (or by email or text) unless you've initiated the call (or communication).	• Prevents identity thieves from obtaining information that can be used to commit fraud.	• Consumers can do online searches to verify identities of requesters, or check with experts, before giving out information. • Consumers should not trust caller ID and should hang up on robocalls and report such calls to FTC at ftc.gov/complaint.
Shred documents and mail with Social Security numbers or other personal information.	• Prevents identity thieves from finding sensitive information in trash.	• Consumers can contact the U.S. Postal Service if they believe their mail is being stolen or misdirected. • Consumers can opt out of receiving credit card and other offers in the mail at 1-888-5-OPT-OUT (1-888-567-8688) or www.optoutprescreen.com.

Source: GAO analysis, Federal Trade Commission, Consumer Financial Protection Bureau, and consumer and industry organizations.

Figure 3. What Can Consumers Do After a Data Breach?

In: Information Security
Editor: Paul Scholz

ISBN: 978-1-53615-880-9

Chapter 8

WHAT LEGAL OBLIGATIONS DO INTERNET COMPANIES HAVE TO PREVENT AND RESPOND TO A DATA BREACH?*

Chris D. Linebaugh

Recently, large Internet companies—i.e., companies that do most of their business on the Internet, such as social media platforms or search engines—have made headlines after failing to secure their users' personal information. For example, on September 28, 2018, Facebook announced a security breach affecting tens of millions of user accounts. According to Facebook, hackers exploited a vulnerability in its code that allowed them to steal "access tokens," which are the "equivalent of digital keys" that "keep people logged in to Facebook." Facebook later disclosed that, of the affected accounts, hackers accessed the names and contact details of 15 million users and the biographical information of another 14 million users. Just over a week after Facebook's breach, on October 8, 2018, Google, in announcing the end of its social network Google+, disclosed that a

* This is an edited, reformatted and augmented version of Congressional Research Service Publication No. LSB10210, dated October 25, 2018.

software glitch exposed the personal data associated with up to 500,000 Google+ accounts. Google explained that it discovered and resolved the glitch in March 2018 and that there was no evidence anyone misused the exposed data. The Internet search giant reportedly made an initial decision not to disclose the incident, before reversing course and shutting down the Google+ platform following a Wall Street Journal investigation.

These incidents raise the question: what legal obligations do Internet companies have to prevent and respond to data breaches? This Sidebar considers the answer to this question under federal and state law. The Sidebar then discusses several factors Congress might consider when weighing future legislation.

FEDERAL AND STATE LAW ON PREVENTING DATA BREACHES

In contrast to the European Union, which recently adopted a wide-reaching data privacy law called the General Data Protection Regulation (GDPR), there is no comprehensive federal law requiring all entities or individuals who collect personal information electronically to maintain the security of such data. Rather, the few federal data security laws that do exist are primarily directed at specific industries. For instance, the Gramm-Leach Bliley Act (GLBA) and its implementing regulations require financial institutions to maintain comprehensive information security programs to safeguard customers' nonpublic personal information. Similarly, the regulations implementing the Health Insurance Portability and Accountability Act (HIPAA) require covered healthcare companies to maintain safeguards to prevent data security threats to electronic "protected health information." Other sector-specific federal laws that require certain entities to adopt data security measures or limit the circumstances under which those entities may disclose personal information include the Driver's Privacy Protection Act (governing state departments of motor vehicles), the Privacy Act of 1974 (governing federal agencies), the Fair

Credit Reporting Act (governing credit reporting agencies), and the Children's Online Privacy Protection Act (COPPA) (governing online operators that direct their services at children or knowingly collect children's data).

Notably, no federal data security law specifically directs Internet companies falling outside the reach of these sector-specific statutes to adopt security measures designed to protect personal data from unauthorized hacks or unintentional exposure. Such companies may nonetheless have obligations under federal consumer protection laws. In particular, the Federal Trade Commission (FTC) Act empowers the FTC to prevent companies from engaging in "unfair or deceptive acts or practices" (commonly referred to as "UDAPs") "in or affecting commerce." The FTC Act allows the FTC to seek equitable relief, including preliminary or permanent injunctions, in UDAP enforcement actions. The Commission may, however, only seek civil monetary penalties if the party has violated a consent decree or a regulation defining a specific type of conduct as a UDAP.

The FTC has used its UDAP authority to bring a number of enforcement actions against companies that fail to invest adequately in data security. Often, the FTC maintains that such failures are "deceptive" because they contradict companies' past security promises. For instance, in 2014 the social media company Snapchat settled FTC charges alleging that its failure to secure its "Find Friends" feature allowed hackers to compile 4.6 million user names and phone numbers. The FTC alleged that Snapchat acted deceptively because the company's privacy policy said that it had taken reasonable data security measures. Along with "deceptive" claims, the FTC has also recognized that failing to safeguard user data adequately may be "unfair." The FTC Act provides that an act or practice is only "unfair" if it "causes or is likely to cause substantial injury to consumers," is not "reasonably avoidable by consumers," and is "not outweighed by countervailing benefits." At least one court has agreed that a company's failure to safeguard user data may, in some circumstances, meet this standard. However, the extent to which the FTC may rely on its "unfairness" authority is unclear. A recent Eleventh Circuit decision suggests that the FTC needs to allege specific failures and remedies in order to successfully issue a cease and desist order based on a company's "unfair" data security measures. In that case, *LabMD v. FTC*, the court noted that the FTC's order "contains no prohibitions" but "commands [the company] to overhaul and replace its data-security program to meet an indeterminable standard of reasonableness." The court concluded that such an order is unenforceable, reasoning that penalizing a company for failing to comply with an imprecise standard "may constitute a denial of due process" and that the order "effectually charges the district court [enforcing the order] with managing the overhaul."

In addition to the FTC Act's UDAP provision, Internet companies must comply with a number of state laws regulating data security. Thus far, at least 22 states have enacted laws of general applicability governing data security. These statutes generally require companies to maintain "reasonable" security procedures to protect personal information. While many state laws do not specify what constitutes a "reasonable" procedure, some states, like Massachusetts, specify in detailed regulations the types of measures required. State consumer protection laws and common law causes of action—such as negligence, negligence per se (often based on federal statutes, like FCRA or GLBA), negligent misrepresentation, fraud, or breach of contract—may also apply. For instance, following Home Depot's data breach affecting more than 50 million cardholders, consumers sued the company alleging negligence, negligence per se, and violations of multiple state consumer protection laws.

FEDERAL AND STATE LAW ON RESPONDING TO DATA BREACHES

Federal laws and regulations impose relatively few obligations on companies that suffer data breaches and are generally directed at specific sectors, such as the banking (under GLBA interagency guidance) or healthcare industries (under HIPAA regulations). However, as the Security and Exchange Commission (SEC) indicated in February 2018 guidance, publicly traded companies may be required to report breaches in their public filings with the SEC because federal securities laws and regulations prohibit companies from omitting material facts necessary to make statements made in connection with the purchase or sale of any security "not misleading." The SEC guidance does not articulate a bright-line approach for determining when companies must report a data breach and does not include any categorical exemptions, such as exemptions for *de minimis* breaches. Rather, the guidance directs companies to consider whether a breach is material to investors in light of factors such as the nature of the compromised information, potential magnitude of the breach, and range of harm caused by the breach. Since issuing the guidance, the SEC has brought at least one enforcement action against a company for failure to report a data breach. In April 2018, Yahoo! Inc.'s successor company agreed to pay a $35 million penalty to settle SEC charges that the company mislead investors by failing to report a major 2014 data breach in which hackers acquired personal data associated with hundreds of millions of user accounts.

Although there are few federal laws regulating responses to data breaches, the states have addressed the issue. As of March 2018, all 50 states have enacted data breach response laws. The laws' specifics vary significantly, however, creating, in the view of some commentators, uncertainty for companies responding to data breaches. For instance, some states require notification after any breach of non-encrypted personal information (e.g., New York), while others require notification only if the breach is likely to cause "substantial harm" to individuals (e.g., Alabama). Furthermore, some states require companies to notify affected individuals within a certain time frame, such as 30 days (e.g., Florida), 45 days (e.g., Washington), or 60 days (e.g., Delaware), while others simply require companies to provide notice "without unreasonable delay" (e.g., California). State laws also differ in how they define personal information, whether and when companies must notify any state agencies, the required contents of the notice, the required method of notice (e.g., some states allow for substitute notice, while others do not), and who can bring actions for violations (while most states only empower the attorney general to enforce violations, some states allow private causes of actions in addition to attorney general enforcement).

Despite their differences, state laws generally require companies to notify affected individuals regardless of whether the breach resulted from an inadequate data security program. State laws typically define a triggering breach as involving the "unauthorized acquisition" of personal information. These definitions generally do not require that the "unauthorized acquisition" be caused by companies' security failures, nor do they necessarily contain carve-outs for companies with adequate data security programs in place (e.g., California, Texas, New York). Because breaches are typically defined as the "unauthorized acquisition" of personal data, companies generally are not required to notify individuals when there is a lack of evidence that the company's failure resulted in a third party possessing personal data without authorization. Nonetheless, a handful of states (Puerto Rico, New Jersey, Florida, and Connecticut) depart from this general rule by defining data breaches to include the "unauthorized access" to personal information. As some commentators have noted, "unauthorized access" may be a lower standard that covers situations where personal data is merely exposed to unauthorized individuals, even if those individuals did not actually gain possession of the data.

Considerations for Congress

While there is a relative absence of federal law on how Internet companies should prevent and respond to data breaches, a number of bills have been introduced on this issue. Any data security legislation raises several legal considerations for Congress. First, Congress might evaluate the scope of the covered subject matter and entities. In terms of subject matter, legislation could impose preventative data breach measures (e.g., H.R. 6864), responsive data breach measures (e.g., H.R. 3816), or both (e.g., H.R. 4081). In terms of entities, legislation could cover all companies subject to the FTC or another agency's jurisdiction (e.g., H.R. 5388), or it could take a sector-specific approach and only cover Internet companies (e.g., S. 2728). Second, Congress might consider the appropriate enforcement agency and the nature of its authority. While many proposed bills would designate the FTC as the primary enforcer, legislation has differed on the way in which the FTC would enforce violations. For instance, some bills would rely on the existing framework under the FTC Act by directing the FTC to treat violations of the bill as violations of a UDAP regulation. Other bills would give the FTC and Department of Justice shared enforcement authority or would create an entirely new "Office of Cybersecurity" to enforce violations. Third, to address concerns about the lack of uniformity among state data breach legislation, Congress might consider the extent to which any new federal legislation would preempt state laws. For example, some legislation would expressly preempt any similar state laws, while others are silent on the issue or would only preempt "less stringent" state laws. These and other legal issues could be of importance as Congress considers legislation in light of recent concerns over how Internet companies are protecting and responding to threats to cybersecurity.

INDEX

A

B

C

D

E

H

I

L

M

N

O

P

R

S

T

U

V

W